MEASUREMENT AND ANALYSIS OF BLAST FRAGMENTATION

WORKSHOP HOSTED BY FRAGBLAST 10 — THE 10TH INTERNATIONAL SYMPOSIUM ON ROCK FRAGMENTATION BY BLASTING, NEW DELHI, INDIA, 24–25 NOVEMBER 2012

Measurement and Analysis of Blast Fragmentation

Editors

José A. Sanchidrián
Universidad Politécnica de Madrid – ETSI Minas, Madrid, Spain

Ashok Kumar Singh
Central Mine Planning & Design Institute Limited, Ranchi, India

CRC Press
Taylor & Francis Group
Boca Raton London New York Leiden

CRC Press is an imprint of the
Taylor & Francis Group, an **informa** business
A BALKEMA BOOK

Published by:
CRC Press/Balkema
Schipholweg 107C, 2316 XC Leiden, The Netherlands

ISBN 13: 978-0-415-62140-3 (hbk + CD-ROM)
ISBN 13: 978-0-203-38753-5 (ebk)

Visit the Taylor & Francis Web site at
http://www.taylorandfrancis.com

and the CRC Press Web site at
http://www.crcpress.com

Typeset by V Publishing Solutions Pvt Ltd., Chennai, India

Measurement and Analysis of Blast Fragmentation – Sanchidrián & Singh (Eds)
© 2013 Taylor & Francis Group, London, ISBN 978-0-415-62140-3

Table of contents

Measurement and Analysis of Blast Fragmentation – Sanchidrián & Singh (Eds)
© 2013 Taylor & Francis Group, London, ISBN 978-0-415-62140-3

Preface

The assessment of fragmentation by blasting and by any of the subsequent crushing and grinding stages is important in order to control and optimize the mining operation. Fragmentation characteristics influence the mucking productivity, crusher throughput and energy consumption, plant efficiency, or the price itself of the end product in the case of industrial minerals and aggregates.

It is clear to everyone that measuring and reporting rock fragmentation means obtaining a size-percentage passing curve of the blasted rock. However, the precise meaning and complications of such size-passing relation are in many cases overlooked. In statistical terms, determining a distribution of sizes involves measuring the sizes of the particles by any suitable gauge system, building a histogram of sizes and deriving a density function of size thereof. Besides this being impossible to carry out with rock samples, usually involving millions of particles, it is not the histogram of numbers, or frequencies, of particles of a given size interval that is of interest, but the amount of material in each size interval. What is meant by amount of material is, in the usual sieving procedure, the mass, or mass fraction, in each interval of mesh size, which represents a discrete density function of the mass for each size. The corresponding representation of the mass fraction of material with size less than a given mesh is a discrete cumulative distribution, the well-known size-fraction, or percentage, passing curve. This is the standard representation of rock fragments size distribution.

Sieving is almost impossible to carry out in a large scale on a permanent basis in a mining operation since it is disruptive of the production work and extremely expensive in time and human resources, though some experiences of full scale sieving exist. In small scale, blasts can be done in a fully controlled way and fragmentation can then be measured accurately by sieving and weighing. Significant experimental efforts to understand the mechanisms of rock fragmentation are being done on the basis of small scale blasting, and two papers on this research area are presented in this volume.

Image analysis techniques have been used for particle size measurement in the last thirty years. As computers and high resolution cameras have become popular and inexpensive, the cost of an image analysis system is relatively low and particle size and shape analyses can be handled easily. For operational control purposes (of blasting, crushing, etc.), sampling and sieving encompass a late feedback of the size distribution data, while image analysis systems can determine the size distribution on-line with reasonable accuracy, and with a set up such that images are taken in the production environment without interfering with the production process. The solutions adopted for image acquisition are very much application dependent but they classically involve 2D camera systems; computational methods involve techniques to elaborate images transformed into numerical information in order to extract edges and delineate (or segment) particles. Segmentation routines are aimed to determine gray intensity or color changes to assign edges or boundaries, this way determining the size and shape parameters of the particles. In this case, usually the size distribution is obtained from the number of particles in size classes, or from the surface covered by them; the amount of material in this case is not mass, but surface, or number of particles. If mass fractions are required, they must be estimated from those by means of experimental conversion factors.

In 1996, on the occasion of Fragblast 5 in Montreal, Canada, a workshop was held that reviewed the state of the art technology on image processing applied to rock fragmentation measurement. It was a brilliant gathering of technical and scientific knowledge and its proceedings volume is, even today, one of the top cited references amongst the blasting community. At that time, image processing was an emerging technology; since then, it has grown and evolved through the years and its use has spread in the mining sector. Along with mathematical research, the development of higher resolution cameras and computers with ever increasing capacity and processing speed have made image analysis a relatively mature technology today, with a number of products available off-the-shelf; image analysis systems are not rare nowadays in mining operations. Difficulties and errors of the image processing techniques are well identified, some of them stemming from the physics of the measurement itself (i.e. trying to measure a property of the bulk of a material by sampling only on its surface, such as segregation, overlapping, capturing errors, etc.)

and others from the unavoidable inaccuracies of the algorithms (e.g. segmentation errors arising from the fact that particles usually have similar texture which makes it difficult to differentiate between particles, and texture or color changes within a single particle that may be wrongly interpreted as boundaries). Methods for measuring the third dimension have been developed by different techniques, such as stereo-photogrammetry, laser triangulation or laser scanning. The leading edge of research in fragmentation measurement by machine vision stands in these 3D systems, of which several solutions are discussed in this volume.

There is no single approach that will solve the fragmentation measurement problem at every site. Practical implementation of image analysis systems include taking images of the muckpile, dipper, truck loads, falling material from trucks while dumping, crusher hoppers and feeders, conveyor belts, etc. A variety of case studies in which fragmentation by blasting is measured and applied to blasting assessment is pictured in several papers hereafter. The combination of measurement and analysis techniques based on Statistics, optimization techniques, on-site calibration and data reconciliation provides an undoubted solution for accurate fragmentation measurement and control. Appropriate functional descriptions of the fragment size, beyond the classical Weibull distribution, may also be instrumental in this task; research in this aspect of fragmentation analysis is also reflected in this volume.

The aim of the workshop has been to bring together researchers and practitioners in the field of measurement and analysis of fragmentation by blasting: developers of measurement systems, users of such systems, along with researchers on fragmentation by blasting performing actual fragmentation measurements by sieving, and also analysts seeking to describe fragmentation by improved distribution functions. Contributions have come from Australia, Austria, Belgium, China, France, India, Iran, Spain and Sweden. This does not make it a comprehensive volume on fragmentation measurement and analysis but it delivers a snapshot of the activity in this field around the world, while the workshop itself provides an opportunity for speakers and attendees to share their experience and their knowledge, to analyze and to discuss the evolution of new techniques and to devise new scenarios where fragmentation measurement techniques can be applied.

We thank the Fragblast International Organizing Committee for the long term support of blasting research and technology worldwide, through the organization of the prestigious Fragblast Symposia. We also thank the National Organizing Committee of this Tenth International Symposium on Rock Fragmentation by Blasting for the initiative of hosting this workshop on fragmentation measurement and analysis, and for providing an excellent support to it throughout. We wish to thank very especially the reviewers for their silent and efficient work and, above all, the authors for taking their time for writing their papers and for sharing their experience and knowledge with all our community.

José A. Sanchidrián
Universidad Politécnica de Madrid – ETSI Minas, Madrid, Spain

Ashok Kumar Singh
Central Mine Planning & Design Institute Limited, Ranchi, India

Measurement and Analysis of Blast Fragmentation – Sanchidrián & Singh (Eds)
© 2013 Taylor & Francis Group, London, ISBN 978-0-415-62140-3

Organising Institution

CSIR - Central Institute of Mining & Fuel Research

(Council of Scientific & Industrial Research)
Dhanbad, India

Measurement and Analysis of Blast Fragmentation – Sanchidrián & Singh (Eds)
 ISBN 978-0-415-62140-3

Committees

PATRONS

- Prof. Samir K. Brahmachari, *Director General, CSIR and Secretary, DSIR, New Delhi*
- Mr. Partho S. Bhattacharyya, *Chairman, CSIR-CIMFR, Research Council, Dhanbad*
- Mr. Satish Puri, *Director General of Mines Safety, DGMS, Dhanbad*
- Mr. S. Narsing Rao, *Chairman-cum-Managing Director, Coal India Limited, Kolkata*

CHAIRMAN ORGANISING COMMITTEE

- Dr. Amalendu Sinha, *Director, CSIR-CIMFR, Dhanbad*

ORGANISING SECRETARY & CONVENOR

- Dr. Pradeep K. Singh, *Senior Principal Scientist, CSIR-CIMFR, Dhanbad*

INTERNATIONAL ORGANISING COMMITTEE

Prof. W.L. Fourney	*University of Maryland, USA*
Prof. José A. Sanchidrián	*Universidad Politecnica de Madrid, Spain*
Docent Agne Rustan	*Retired from Luleå University of Technology, Sweden*
Prof. Hans Peter Rossmanith	*Technical University, Vienna, Austria*
Prof. Sushil Bhandari	*Earth Resource Technology Consultants, India*
Dr. Cameron K. McKenzie	*Blastechnology, Australia*
Prof. Bibhu Mohanty	*University of Toronto, Canada*
Prof. Xuguang Wang	*Beijing General Research Institute of Mining & Metallurgy, China*
Mr. R. Frank Chiappetta	*Blasting Analysis International, USA*
Mr. Carlos P. Orlandi	*Enaex Servicios S.A., Chile*
Prof. Finn Ouchterlony	*Montanuniversitaet Leoben, Austria*
Prof. Kunihisa Katsuyama	*(Retired from) Ehime University, Japan*
Dr. William Robert Adamson	*Davey Bickford, Chile*
Prof. Panagiotis D. Katsabanis	*Queen's University, Canada*
Prof. Peter Moser	*Montanuniversitaet Leoben, Austria*
Dr. Ken Qian Liu	*Xstrata Nickel, Canada*
Dr. Ewan Sellers	*African Explosives, South Africa*
Dr. A.T. Spathis	*Orica, Australia*
Dr. Dale S. Preece	*Orica Mining Services, USA*

REVIEW COMMITTEE

Prof. W.L. Fourney	*University of Maryland, USA*
Prof. José A. Sanchidrián	*Universidad Politecnica de Madrid, Spain*
Docent Agne Rustan	*Retired from Luleå University of Technology, Sweden*
Prof. Sushil Bhandari	*Earth Resource Technology Consultants, India*

Sponsors

Diamond Sponsor
Coal India Limited

Platinum Sponsors
Solar Industries India Limited

Orica

Deepak Fertilisers & Petrochemicals
Corporation Ltd.

Hindustan Zinc Limited
(Vedanta Group Company)

Tata Steel

Gold Sponsors

Singareni Collieries Company Limited

Jindal Steel & Power Limited

Silver Sponsors

NMDC Ltd.

Essel Mining & Industries Limited

National Aluminium Company Limited

Jaiprakash Industries Limited

Manganese Ore (India) Limited

Uranium Corporation of India Limited

EMTA Group of Companies

Sarda Mines Pvt. Ltd.

Bronze Sponsors

Gujarat Mineral Development Corporation

IDL Explosives Limited

J.K. Cement

Hutti Gold Mines Limited

V.V. Mineral (VVM)

Navbharat Group of Companies

JSW Bengal Steel Ltd.

Neyveli Lignite Corporation Limited

Lunch Sponsor

Ganesh Explosive Pvt. Ltd.

Technology developments

Measurement and Analysis of Blast Fragmentation – Sanchidrián & Singh (Eds)
© 2013 Taylor & Francis Group, London, ISBN 978-0-415-62140-3

A history of digital image analysis technique for blast fragmentation assessment and some Indian contributions

A.K. Raina
Central Institute of Mining & Fuel Research, Nagpur, India

ABSTRACT: Blast fragmentation assessment is fundamental to the optimization of Mine-Mill-Fragmentation System (MMFS). In absence of a better method of fragmentation assessment, digital image analysis technique has assumed significance. The journey of such software has seen manual to semi-automatic to fully automatic transformations. Despite of the excellent efforts by research and programming groups some inherent problems still pose problems in estimations and statistical accuracies of such software. The underlying parameters of quality of the images, the handling capacity of such systems vis-à-vis the time of analysis, 2D to 3D conversion techniques etc. provide ample space for critics to downplay the importance of such systems. Despite of the critique, the software are in vogue for more than 25 years now. There is sufficient evidence to document the use of such software but the problems faced by the user communities are least reported. This paper tries to documents the chronology of development and put a word for future requirements and direction of development of such systems.

1 INTRODUCTION

Accurate measurement of blast fragmentation is fundamental to the productivity in mines for simple reason that this parameter directly influences the performance of the mechanical equipment deployed for loading and hauling. Digital Image Analysis Technique (DIAT) is one of the recognized methods for determination of fragmentation distributions resulting from blasting. A good number of attempts by the research community have resulted in evolution of different software systems out of which few have been able to sustain in the competitive arena. The basic premise for any software is to compete with the developments in other fronts and same applies to DIAT also. Since the audience for such software is miniscule in size, the constant upgradation is practically a difficult proposition.

Despite this, DIAT is a method in vogue for more than 25 years now. These systems can be broadly classified into two groups:

1. Static systems—in which images of blasted muck are collected and analyzed after collecting post blast images.
2. Dynamic—these systems are also called online systems and are generally fixed over conveyors belts and capture and analyze crusher/plant feed fragment sizes at defined intervals. See Wang & Stephansson (1996) and Downs & Kettunen (1996).

A comprehensive list of such software is presented by Chakraborty et al. (2002; Table 1) and a treatise on the subject (Franklin & Katsabanis 1996) provides a host of information about the subject.

The DIAT is well received by the industry despite several challenges (Cunningham 1996a) arising out of the technique being utilized for assessment of fragment sizes. These include:

1. *The representative sampling size* or the number of images analyzed. Since the volume of the blasted rock may range in thousands of tons, the exact number of images that qualify being representative is difficult to determine. Statistically, better the sample size better the accuracy holds well but it can have its financial implications in the long run. The techniques for the sampling are given by Maerz (1996b).
2. *The quality of images that are put to analysis.* Modern day cameras have high resolution and higher image sizes. However, the software used for analysis are not tuned or capable to synthesize the fragment information, do not have capability to handle such high resolution images and may take lot of time to analyze such images. This raises a question about the compromise of quality of results attained from the images. Some of the relevant issues are attended by Palangio & Franklin (1996).
3. *The scaling errors arising from tilted images* since an aspect error always exists in the images. Most of the software have the capability to adjust the tilt during analysis or in real-time.
4. *The errors due to wrong/poor edge detection* techniques (see Bedair et al. 1996). Since most

Table 1. Various digital analysis software globally available and the capabilities (Chakraborty et al. 2002).

Image processing system (Acronym or shortened form)	Manual editing	Number of classes (bins)	Accuracy (%)
CIAS (USA), USBM, & Noramco Engg.	No	6	10–20
FRAGSCAN (France). Ecole des Mines de Paris	No	8–15	~10
Gold size (USA and Canada), Golder Associates	Yes	100	–
IPACs (Sweden)	No	–	–
KTH (Sweden)	No	–	–
Power Sieve (Australia)	Yes	User defined	–
Split (USA)	Yes	<100	~10
TUCIPS (Germany)			~5
Fragalyst$^{3.0}$ (India)	Yes	5	~10
WipFrag (Canada)	Yes	40	2–16

of the systems work on the greyscale method of edge detection, the possibility of identifying the edges wrongly exists.

5. *Errors arising out of bias towards the fines*. This is due to the fact that images have their own resolution and the size of fines is less than the size a DIAT system can resolve. Software rely on fines correction schemes developed through sieve analysis calibrations to correct this.
6. *2D to 3D unfolding models*—a lot of published works exist on the subject (Maerz 1996a) and probably the same is not resolved as yet as there is a difference of opinion.
7. *Errors due to distribution fitting*. These arise once a particular distribution is fitted to the measured fragmentation data. If the fit is not good there is a possibility of wrong prediction of the mean fragment size and uniformity index. Recent works of Ouchterlony (2005, 2010), has brought out this problem explicitly and proposed new functions to account for loss of fines due to fitting of R-R distribution.

2 HISTORICAL DEVELOPMENTS

2.1 *World scenario*

A historical perspective of the developments world over and use of DIAT is summarized in Table 2.

2.2 *Indian scenario*

Software products have also been used in India to determine the blast fragment sizes. One such program is Fragalyst (developed by CSIR-CIMFR and Wavelet Group). The relevant published studies indicating use of Fragalyst and citations are given in Table 3.

Some features of the Fragalyst were detailed by Raina et al. (2010). The features in the latest version are introduced below.

2.2.1 *Features in latest version of fragalyst*

Fragalyst is a windows based fragmentation analysis system that has many features which satisfy the objectives of the software. The journey of the software dates back to 1997 and since then the software has seen several version changes and improvements.

2.2.2 *Objective*

The software has been designed to work out the effective energy utilization in blasting for which the in situ block size distribution can be determined by 2 methods

1. Manual input of the joint spacing, or
2. Input from images of the faces, this method works out the spacing and length of joints and calculates the in-situ block size distribution.

2.2.3 *Calibration*

Two calibration modes viz. Normal and Tilt correction are available in the software. The effectiveness of the tilt correction is demonstrated in Figure 1. If an image is not properly calibrated, it will get distorted and is thus a measure to check whether proper calibration is enacted or not.

Another feature with the calibration is zoomed edges of the calibrator for exact measurement of the scale.

2.2.4 *Image handling and enhancement*

The software has capability to handle large image sizes of more than 3MP which improves the accuracy of the results. Other features of image handling are:

1. Image enhancement—brightness, contrast and sharpness,
2. Image size—reduction or enlargement options,
3. Crop facilities, etc.

Table 2. Important published literature on developments and use of DIAT.

Year	Author	Description with respect to DIAT
1987	Maerz et al.	New DIAT to measure the blasted block size, tested with real-time monitoring.
1990	Hendricks & Scoble	Fragmentation, shovel performance
1991	Hendricks & Scoble	Fragmentation
1991	Scoble et al.	Fragmentation with WIEP, equipment performance, downstream effects
1992	Chung & Ludwig	Compaphoto, Fragmentation, drawbacks
1994	Eloranta	Fragmentation (USBM program), effect of stemming
1995	Hanspal et al.	Anatomy of a blast muckpile, fragmentation sizes
1995	Franklin et al.	WipFrag, quality control of underground blasting operations, key features of the granulometry, statistical choices
1995	Palangio	WipFrag (Franklin, Geotechnical Ltd., ETI Explosives, INCO and the University of Waterloo)
1996b	Cunningham	Challenges of DIAT & Compaphoto technique
1996a	Maerz	2-D to 3-D block size reconstruction
1996	Franklin et al.	Evolution of measurement systems
1996	Dahlhielm	The IPACS system
1996	Havermann & Vogt	TUCIPS
1996	Schleifer & Tessier	FRAGSCAN
1996	Downs & Kettunnen	CIAS (Online measurement)
1996	Kleine & Cameron	GoldSize
1996	Maerz et al.	WipFrag—granulometry system
1996	Girdner et al.	Split System
1997	Michaud et al.	Development, pro and cons of the DIAT, sampling method
1997	Eloranta	Energy requirements for blasting, crushing and grinding, fragmentation
1997	Kleine & Cameron	Methodology for DIAT, GoldSize use
1997	Noy	Validity of 2D to 3D unfolding transformation, standards in fragmentation measurement in 2D analysis
1998	Bozic	Fragmentation (DIAT), need for a new generation of engineering tools
1999	Barkley & Carter	WipFrag use, precision and measurements in DIAT (reproducibility, sample size, optical resolution and problem of 'lost fines')
1999	Katsabanis	DIAT vs. screening on laboratory scale, fines bias
1999	Higgins et al.	Split use, blast optimization, pre-, post blast fragmentation, sampling
2000	Sarma	"Mine-Mill Blasting" approach, simulations, case histories, Split use
2000	Lin et al.	Online Particle Size Analyser (OPSA)
2001	Eloranta	Fragmentation optimization through DIAT
2001	Grundstrom et al.	Fragmentation optimization, using Split, significant improvements in sag mill throughput
2002	Hendricks	Design and application of AQUILA's drill systems, Fragmentation WipFrag use
2002	Cho et al.	Numerical simulation, fragmentation prediction, fracture mechanism
2002	Mc Kinstry et al.	DIAT, electronic detonators use
2002	Wardrip et al.	Performance of a high density, high velocity cartridged explosive
2003	Mckinstry et al.	DIAT, electronic detonators use
2003	Workman	Blasting energy input, fragmentation, feed sizes effect on energy consumption in crushing and grinding
2003	Eloranta	DIAT system, tracked 3 three key size fractions within the resolvable range of the lens, calibration errors
2003	Latham et al.	FragScan, PowerSieve, Split and WipFrag blind comparison, strengths and weaknesses of DIAT
2004	Hosseini et al.	GoldSize use, fragment size for a proposed conveyor
2004	McKinstry et al.	DIAT, gains in mill throughput.
2005	Kay & Song	Fragmentation, tunnelling, electronic detonators

(*Continued*)

Table 2. Continued.

Year	Author	Description with respect to DIAT
2005	Kahriman et al.	DIAT, Kuz-Ram model, comparison of results
2005	Glowe	DIAT, results good for individual parts of a blast
2006	Singh et al.	Disparity in sieved and DIAT measured results, fines bias
2006	Sanchidrian et al.	DIAT, fragmentation measurement, procedures
2006	Potts & Ouchterlony	Capacity of DIAT, fragmentation
2006	Kim	Specific explosives energy, fragmentation reduction ratio, rock fracture toughness, DIAT use
2006	Ozkahraman	Fragmentation, blast design DIAT
2007	Özdemir et al.	Split use, particle size distribution
2007	Chavez et al.	MMFS optimization, DIAT
2007	Maki	DIAT, fragmentation, electronic detonators
2007	Eloranta et al.	History of DIAT, current static and on-line analysis technology
2007	Al-Thyabat et al.	Moving particles (online systems), problems, DIAT
2008	King	Use of cameras by the blasting industry, DIAT
2008	Palangio	DIAT, equipment, methods, techniques and information, developments
2008	Kahriman et al.	DIAT, fragmentation, cycle time relations of equipment
2008	Outal et al.	Calibration method 3D distribution from 2D images
2009	Singh et al.	Methods, operating procedures, problems, limitations of DIAT (CS)
2009	Koh et al.	Hardware based approach, segmentation, multi-flash imaging
2009	Sanchidrian et al.	Accuracy of DIAT
2010	Paley	Split use, fragmentation, SAG mill performance electronic detonators
2010	Quezada et al.	Effect of size of the fines on product, DIAT use
2010	Outal et al.	Calibration method 3D distribution from 2D images
2010	Sanchidrian et al.	Distribution functions evaluation
2010	Thurley	Fragmentation measurement using 3D surface imaging
2010	Engin	Blast design based on DIAT
2010	Raina et al.	Productivity improvement by DIAT
2012	Jemwa & Aldrich	Machine vision and elements from statistical pattern recognition, support vector machine

Table 3. Use of Fragalyst in cited works.

Year	Author	Summary of citation to Fragalyst
2000	Sarathy	Fragmentation issues
2001	Ramulu et al.	Blast fragmentation measurement and improvements
2002	Chakraborty et al.	Blast fragmentation measurement and development of models for fragmentation
2004	Chakraborty et al.	Parametric study, blast optimization
2002	Bagde et al.	Rock characterization
2002	Raina et al.	Fragalyst introduction
2003	Pal Roy et al.	Fragmentation, induced caving
2003	Thote & Singh	Fragmentation optimization
2004	Chakraborty et al.	Blast optimization
2004	Rai & Singh	Initiation sequence, blast fragmentation
2005	Rai et al.	Fragmentation optimization
2006	Sudhakar et al.	Comparison of fragmentation by different software
2008	Rai et al.	Fragmentation
2010	Raina et al.	Fragalyst application and analysis
2011	Yang & Rai	Fragmentation and delay timing
2012	Rai et al.	Slot cut development and fragmentation

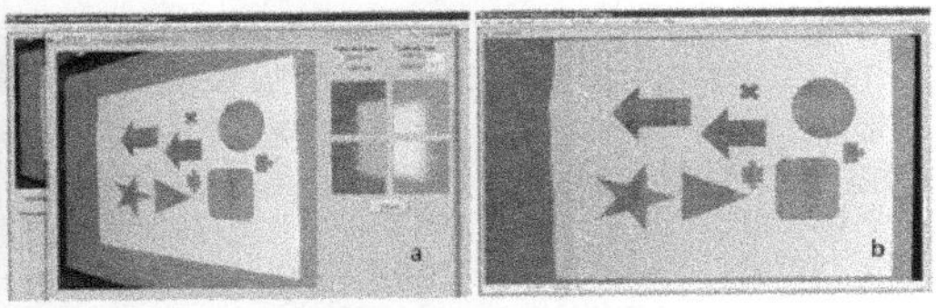

Figure 1. Tilt correction in Fragalyst; a) actual, b) corrected image.

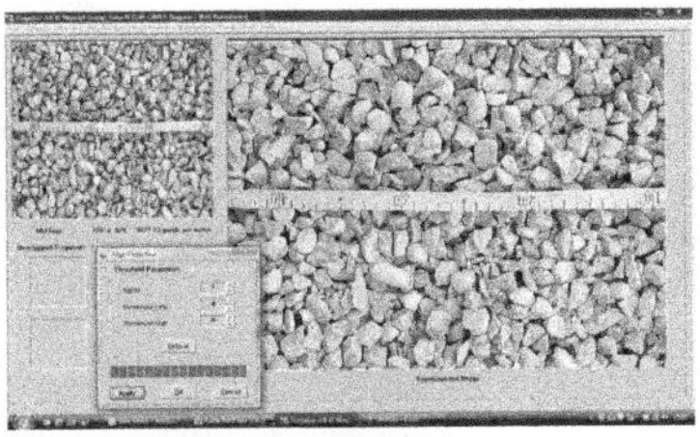

Figure 2. Edge detection in auto-mode.

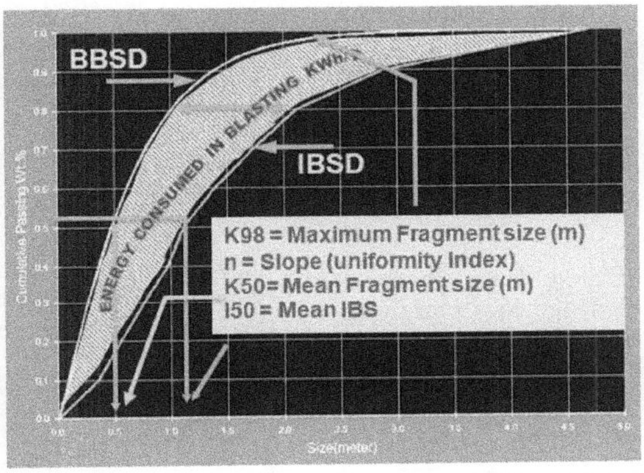

Figure 3. Evaluation of effective energy utilization from IBSD & BBSD.

Table 4. Technical comparison of different generations of fragmentation assessment systems and future requirements.

	Generation			
Specifications	1st	2nd	3rd	4th (Future)
Image capture facility	Scanner	Camera	Camera	Laser, wave reflection
Image zoom	N	Y	Y	NR
Image enhancement	N	Y	Y	NR
Image handling	Difficult	Better	Easy	NR
Lighting	Y	Y	Y	NR
Sample limit	Few	10–15	20	Any
Calibration aspect ratio	N	Y	Y	Automatic
Image size	<1 MP	1–2 MP	>2 MP	Any
IBSD	N	Y stats	Y stats	Y intelligent
Edge detection	Manual	Quasi-automatic	Semi-automatic	Automatic
Network editing	High	Moderate	Low	NR
Fines correction	N	N	Y	Y
Basis of measurement	Relative	Relative	Relative	Absolute
Merge file options	N	Y	Y	Database
IBSD/BBSD comparison	N	N	Y	Intelligent
Energy utilization	N	N	Y	Intelligent
Data handling	N	N	N	Relational
Blast design based on energy utilization	N	N	N	Possible
Online analysis	N	N	N	Possible
Cloud computing	N	N	N	Possible

2.2.5 *Edge detection*

The principle of the quality of images applies to this software also. The edge detection routine in this software is based on an intelligent mechanism that determines the best edge detection parameters from the image and greyscale parameters which can be manually changed, also. A representative image of detected edges is shown in Figure 2.

2.2.6 *In-situ block size determination from image and effective energy utilization*

In situ block size can be estimated from calibrated images of a blast face. The in situ block size and blasted block size distributions can be plotted in the software to get the effective energy utilization as shown in Figure 3.

There are several other features in the software like:

1. Fines correction,
2. Easy data handling,
3. Normal distributions of area, diameter, sphe-rocity, shape factor of the fragments analyzed,
4. Mensuration—Length and angle measurements
5. Simple blast design, and
6. One touch help

3 FUTURE OF FRAGMENTATION ANALYSIS

DIAT has seen a sea change over a period of 25 years with its use in many mines and simultaneous upgradation of different such systems in place. A comparison of the developments has been given in Table 4 in order to make an appraisal of past, present and the future requirement of the fragmentation assessment.

The modern day technologies can provide excellent solutions to the problem of fragmentation assessment. There is a possibility of developing systems which measure fragmentation in real time and assess the same over a wide span of time and mass. A basis for such technology is presented by Thurley (2010) and several others are working on different aspects of the problem to eliminate several shortcomings of the existing system. In addition, intelligent systems can also be devised which is a need of the hour. The excellent processing power of computers can be used to such advantage. The determination of both in-situ and blasted fragment size will have to be taken into consideration for such systems.

4 CONCLUSION

Digital image analysis for fragmentation evaluation has been in vogue for 25 years now and has seen lot of changes in the methodology and techniques being used for the same. There is good number of references in the published domain to support the usefulness of the DIAT. The future of the technique will dwell on different technologies in isolation or in conjunction to deliver products which have capabilities of real-time monitoring and yielding the absolute values of the fragmentation over an unlimited sample size. However, the development of such system will be constrained by its acceptance by the industry and readiness to fund R&D schemes on this important aspect of mining. A thorough introspection and interaction between scientists and developers can address several issues and present a plausible solution for the future of the DIAT.

ACKNOWLEDGEMENTS

The authors are thankful to D'CIMFR for his permission to publish the paper. Help rendered by Dr. P. Pal Roy and other colleagues is gratefully acknowledged.

REFERENCES

Al-Thyabat, S., Miles, N.J. & Koh, T.S. 2007. Estimation of the size distribution of particles moving on a conveyor belt. *Minerals Engineering* 20: 72–83.

Bagde, M.N., Raina, A.K., Chakraborty, A.K. & Jethwa, J.L. 2002. Rock mass characterization by fractal dimension. *Engineering Geology* 63(1–2): 141–155.

Barkley, T. & Carter, R. 1999. Evaluation of optical sizing methods. *Gen. Proc. 29th Ann. Conf. of ISEE on Expl. & Blasting Tech., Nashville, Feb. 2–5*, pp. 305–323.

Bedair, A., Daneshmend, L.K. & Hendricks, C.F.B. 1996. Comparative performance of a novel automated technique for identification of muck pile fragment boundaries. In J.A. Franklin & T. Katsabanis (eds.), *Measurement of Blast Fragmentation*. Rotterdam, Balkema, pp.157–166.

Bozic, B. 1998. Control of fragmentation by blasting. *Rudarsko-geoloiko-nafini zbornik* 10: 49–57.

Chakraborty, A.K., Raina, A.K., Ramulu, M, Choudhury, P.B., Dudhankar, A., Jhanwar, J.C. & Jethwa, J.L. 2002. Development of innovative models for optimisation of blast fragmentation and muck profile applying image analysis technique and subsystems utilisation concept in Indian surface coal mining regime. Project Report, MOC Govt. of India Funded Research Project.

Chakraborty, A.K., Raina, A.K., Ramulu, M., Choudhury, P.B., Haldar, A. Sahu, P. & Bandopadhyay, C. 2004. Parametric study to develop guidelines for blast fragmentation improvement in jointed and massive formations. *Engineering Geology* 73(1–2): 105–116.

Chavez, R., Leclercq, F. & McClure, R. 2007. Applying up-to-date blasting technology and Mine to

Mill concept in quarries. *Gen. Proc. 33rd Ann. Conf. of ISEE on Expl. & Blasting Tech., Jan. 28–31, Nashville,* 1, pp. 329–341.
Cho, S.H., Nishi, M., Yamamoto, M., Kato, M. & Katushiko, K. 2002. Estimation of rock fragmentation in bench blasting using numerical simulation. *Gen. Proc. of the 28th Ann. Conf. on Expl. and Blasting Tech., Feb. 10–13, Las Vegas,* 1, pp. 187–196.
Chung, S.H. & Ludwig, G. 1992. Semi-automated fragmentation assessment. *Res. Proc. 18th Ann. Conf. of ISEE on Expl. & Blasting Tech., Jan. 19–23, Orlando*, pp. 131–138.
Cunningham, C.V.B. 1996a. Optical fragmentation assessment—A technical challenge. In J.A. Franklin & T. Katsabanis (eds.), *Measurement of Blast Fragmentation*. Rotterdam, Balkema, pp. 13–19.
Cunningham, C.V.B. 1996b. Lessons from the Compaphoto technique of fragmentation measurement. In J.A. Franklin & T. Katsabanis (eds.), *Measurement of Blast Fragmentation*. Rotterdam, Balkema, pp. 53–58.
Dahlhielm, S. 1996. Industrial applications of image analysis—the IPACS system. In J.A. Franklin & T. Katsabanis (eds.), *Measurement of Blast Fragmentation*. Rotterdam, Balkema, pp. 59–66.
Downs, D.C. & Kettunen, B.E. 1996. On-line fragmentation measurement utilizing the CIAS system. In J.A. Franklin & T. Katsabanis (eds.), *Measurement of Blast Fragmentation*. Rotterdam, Balkema, pp. 79–82.
Eloranta, J. 1994. Stemming selection for large-diameter blastholes. *Gen. Proc. of the 20th Ann. Conf. on Expl. & Blasting Tech. Jan. 30–Feb. 3, Austin*, pp. 255–266.
Eloranta, J. 1997. The efficiency of blasting verses crushing and grinding. *Gen. Proc. of the 23rd Ann. Conf. on Expl. & Blasting Tech., Feb. 2–5, LasVegas*, pp. 157–163.
Eloranta, J. 2001. Improve Milling Through Better Powder Distribution. *Gen. Proc. of the 27th Ann. Conf. on Expl. & Blasting Tech., Jan. 28–31, Orlando,* 1, pp. 55–63.
Eloranta, J. 2003. Characterization of the pre and post blast environments, *Gen. Proc. of the 29th Ann. Conf. on Expl. and Blasting Tech., Feb. 2–5, Nashville,* 2, pp. 263–264.
Eloranta, J., Palangio, T., Palangio, T.C. & Workman, L. 2007. Size Matters on the Mesabi Range. *Gen. Proc. 33rd Ann. Conf. of ISEE on Expl.& Blasting Tech., Jan. 28–31, Nashville,* 1, pp. 387–399.
Engin, I.C. 2010. A practical method of bench blasting design for desired fragmentation based on digital image processing technique and Kuz-Ram model. In J.A. Sanchidrian (ed.), *Fragmentation by Blasting, Proc. of the 9th Int. Symp. on Rock Fragmentation by Blasting—Fragblast 9, Sept. 13–17, 2009, Granada Spain*, pp. 257–264.
Franklin, J.A. & Katsabanis, T. 1996. Measurement of Blast Fragmentation. Rotterdam: Balkema.
Franklin, J.A., Kemeny, J.A. & Girdner, K.K. 1996. Evolution of measurement systems: A review. In J.A. Franklin & T. Katsabanis (eds.), *Measurement of Blast Fragmentation*. Rotterdam, Balkema, pp. 47–52.
Franklin, J.A., Maerz, N.H. & Santamarina, J.C. 1995. Developments in blast fragmentation measurement. *Res. Proc. of the 21st Ann. Conf. on Expl. and Blasting Tech., Feb. 5–9, Nashville,* 1, pp. 258–269.
Girdner, K.K., Kemeny, J.M., Srikant, A. & McGill, R. 1996. The split system for analysing the size distribution of fragmented rock. In J.A. Franklin & T. Katsabanis (eds.), *Measurement of Blast Fragmentation.* Rotterdam, Balkema, pp. 101–110.
Glowe, R. 2005. Blasting results compared using crusher power consumption and tonnage of rock produced. *Gen. Proc. of the 31st Ann. Conf. on Expl. and Blasting Tech., Feb. 6–9, Orlando,* 2, pp. 1–15.
Grundstrom, C., Sarma, S.K., Jankovich, A. & Thornton, D. 2001. Blast Fragmentation for Maximising the Sag Mill Throughput at Porgera Gold Mine. *Gen. Proc. of the 27th Ann. Conf. on Expl. and Blasting Tech., Jan. 28–31, Orlando* 1, pp. 383–399.
Hanspal, S., Scoble, M. and Lizotte, Y., 1995, Anatomy of a blast muckpile and its influence on loading machine performance. *21st Conf. on Expl. and Blasting Tech., ISEE, Feb. 5–9, Nashville,* 1, pp. 57–67.
Havermann, T. & Vogt. W. 1996. TUCIPS—A system for estimation of fragmentation after production blasts. In J.A. Franklin & T. Katsabanis (eds.), *Measurement of Blast Fragmentation.* Rotterdam, Balkema, pp. 67–72.
Hendricks C. 2002. The Application of GPS Positioning and Drill Monitoring Technology. *Gen. Proc. of the 28th Ann. Conf. on Expl. and Blasting Tech., Feb. 10–13, Las Vegas,* 1, pp. 69–81.
Hendricks, C. & Scoble, M. 1990. Post-blast evaluation through shovel performance monitoring. *Gen. Proc. of the 16th Ann. Conf. on Expl. and Blasting Tech., Feb. 4–9, Orlando*, pp. 227–238.
Hendricks, C. & Scoble, M. 1991. Blast assessment through shovel performance monitoring. 3rd High-Tech Seminar Blasting Technology, Instrumentation and Explosives Applications, San Diego, California, June 2–7, Section 23, 48p.
Higgins, M., BoBo, T., Girdner, K., Kemeny, J. & Seppala, V. 1999. Integrated software tools and methodology for optimization of blast fragmentation. *Gen. Proc. of the 25th Ann. Conf. on Expl. and Blasting Tech., Feb. 7–10, Nashville,* 2, pp. 355–368.
Hosseini, Z., Amirrahmat, M. & Hassan Khoshroo, S. 2004. Determination of Ore Fragmentation by Image Analysis in Sarcheshmeh Mine. *Gen. Proc. of the 30th Ann. Conf. on Expl. and Blasting Tech., Feb 1–4, New Orleans,* 1, pp. 99–107.
Jemwa, G.T & Aldrich, C. 2012. Estimating size fraction categories of coal particles on conveyor belts using image texture modeling methods. *Expert Systems with Applications* 39: 7947–7960.
Kahriman, A., Özer, U., Karadoğan, A., Özdemir, K., Tuncer, G. & Kaya, E. 2008. Effects of particle size distribution on loading performance, *Gen. Proc. of the 34th Ann. Conf. on Expl. and Blasting Tech., Jan. 27–30, New Orleans,* 1, 6p.
Kahriman, A., Tuncer, G., Karadogan, A. & Ozdemir, K. 2005. Kuz-Ram and digital image processing system combination to determine specific blasting parameters. *Gen. Proc. of the 31st Ann. Conf. on Expl. and Blasting Tech., Feb. 6–9, Orlando,* 1, pp. 399–205.
Katsabanis, P.D. 1999. Comparison between image analysis and sieving using laboratory scale model muckpiles. *Gen. Proc. of the 25th Ann. Conf. on Expl. and Blasting Tech., Feb. 7–10, Nashville,* 2, pp. 325–332.

Kay, D. & Song, Y-S. 2005. Civil tunnelling with electronic blasting systems—early experience from Korea. *Gen. Proc. of the 31st Ann. Conf. on Expl. and Blasting Tech., Feb. 6–9, Orlando,* 1, pp. 147–158.

Kim, K. 2006. Blasting Design Using Fracture Toughness and Image Analysis of the Bench Face and Muckpile. MS Thesis, Virginia Polytechnic Institute and State University, 137p.

King, R. 2008. Utilize photography as evidence. *Gen. Proc. of the 34th Ann. Conf. on Expl. and Blasting Tech., Jan. 27–30, New Orleans,* 1, 15p.

Kleine, T.H. & Cameron, A.R. 1996. Blast fragmentation measurement using GoldSize. In J.A. Franklin & T. Katsabanis (eds.), *Measurement of Blast Fragmentation*. Rotterdam, Balkema, pp. 83–90.

Kleine, T.H. & Cameron, A.R. 1997. A blast fragmentation measurement and prediction system for blast optimization *Res. Proc. of the 23rd Ann. Conf. on Expl. and Blasting Tech., Feb. 2–5, Las Vegas*, pp. 89–99.

Koh, T.K., Miles, N.J., Morgan, S.P. & Hayes-Gill, B.R. 2009. Improving particle size measurement using multi-flash imaging. *Minerals Engineering* 22: 537–543.

Latham JP, Kemeny J, Maerz N, Noy M, Schleifer J, Tose S. 2003. A blind comparison between results of four image analysis systems using a photo-library of piles of sieved fragments. *Fragblast J.* 7(2): 105–32.

Lin, C.L., Yen, Y.K. & Miller, J.D. 2000. Plant-site evaluations of the OPSA system for on-line particle size measurement from moving belt conveyors. *Minerals Engineering* 13(8–9): 897–909.

Maerz, N.H. 1996a. Reconstructing 3-D block size distributions from 2-D to measurements on sections. In J.A. Franklin & T. Katsabanis (eds.), *Measurement of Blast Fragmentation*. Rotterdam, Balkema, pp. 39–46.

Maerz, N.H. 1996b. Image sampling techniques & requirements for automated image analysis of rock fragmentation. In J.A. Franklin & T. Katsabanis (eds.), *Measurement of Blast Fragmentation*. Rotterdam, Balkema, pp. 115–120.

Maerz, N.H., Franklin, J.A., Rothenburg, L. & Coursen, D.L. 1987. Measurement of rock fragmentation by digital photoanalysis, *Proc. 6th Intern. Cong. Int. Soc. Rock Mechanics*, 1, pp. 687–692.

Maerz, N.H., Palangio, T.C. & Franklin, J.A. 1996. WipFrag image based granulometry system. In J.A. Franklin & T. Katsabanis (eds.), *Measurement of Blast Fragmentation*. Rotterdam, Balkema, pp. 91–100.

Maki, K. 2007. Digital Detonators in Iron Ore. *Gen. Proc. of the 33rd Ann. Conf. on Expl. & Blasting Tech., Jan. 28–31, Nashville*, 1, pp. 365–371.

Mc Kinstry, R., Floyd, J. & Bartley, D. 2003. Electronic Detonator Performance Evaluation. *The Journal of Explosives Engineering* 19(3): 12–21.

McKinstry, R., Bolles, T. & Rantapaa, M. 2004. Implementation of Electronic Detonators at Barrick Goldstrike Mines, Inc. *Gen. Proc. of the 30th Ann. Conf. on Expl. and Blasting Tech., Feb. 1–4, New Orleans,* 1, pp. 349–361.

McKinstry, R., Floyd, J., Bartley, D. 2002. Electronic detonator performance evaluation, Barrick Goldstrike Mines Inc. *Gen. Proc. of the 28th Ann. Conf. on Expl. and Blasting Tech., Feb. 10–13, Las Vegas,* 2, pp. 1–20.

Michaud, P., Lizotte, Y. & Scoble, M. 1997. Rock fragmentation and mining productivity: characterization and case studies. *Gen. Proc. of the 23rd Ann. Conf. on Expl. and Blasting Tech., Feb. 2–5, Las Vegas: 61–72.*

Noy, M. 1997. 2D versus 3D fragmentation analysis: preliminary findings. *Res. Proc. of the 23rd Ann. Conf. on Expl. and Blasting Tech., Feb. 2–5, Las Vegas*, pp. 181–189.

Ouchterlony, F. 2005. The Swebrec function: Linking fragmentation by blasting & crushing. *Min. Tech. (Trans. Inst. Min. Metall.)* 114: A29–A43.

Ouchterlony, F. 2010. Fragmentation characteristics; the Swebrec function and its use in blast engineering. In J.A. Sanchidrian (ed.), *Fragmentation by Blasting, Proc. of the 9th Int. Symp. on Rock Fragmentation by Blasting—Fragblast 9, Sept. 13–17, 2009, Granada Spain*, pp. 3–22.

Outal, S., Jeulin, D. & Schleifer, J. 2008. A new method for estimating the 3-D size- distribution-curve of fragmented rocks out of 2D images. *Image Anal Stereol* 27: 97–105.

Outal, S., Schleifer, J. & Pirard, E. 2010. Evaluating a calibration method for the estimation of fragmented rock 3D-size-distribution out of 2D images. In J.A. Sanchidrian (ed.), *Fragmentation by Blasting, Proc. of the 9th Int. Symp. on Rock Fragmentation by Blasting—Fragblast 9, Sept. 13–17, 2009, Granada Spain*, pp. 221–228.

Özdemir, K., Tuncer, G., Kahriman, A., Özer, Ü. & Karadoğan, A. 2007. The relation between excavator bucket loading time and particle size distribution of shot rock. *Gen. Proc. of the 33rd Ann. Conf. on Expl. and Blasting Tech., Jan. 27–30, Nashville,* 1, pp. 303–314.

Ozkahraman, H.T. 2006. Fragmentation assessment and design of blast pattern at Goltas Limestone Quarry, Turkey. *Int. J. Rock Mech. Min. Sci.* 43: 628–633.

Pal Roy, P., Sawmliana, C., Bhagat, N.K. & Madhu, M. 2003. Induced caving by blasting: innovative experiments in blasting gallery panels of underground coal mines of India. *Mining Technology* 112(1): 57–63(7).

Palangio, T.C. & Franklin, J.A. 1996. Practical guidelines for lighting and photography. In J.A. Franklin & T. Katsabanis (eds.), *Measurement of Blast Fragmentation*. Rotterdam, Balkema, pp. 111–114.

Palangio, T.C. 1995. WIPFRAG—a new tool for blast evaluation. *Res. Proc. of the 21st Ann. Conf. on Expl. and Blasting Tech., Feb. 5–9, Nashville*, pp. 269–280.

Palangio, T.C. 2008. Digital image analysis. *Gen. Proc. of the 34th Ann. Conf. on Expl. and Blasting Tech., Jan. 27–30, New Orleans*, 1, 10p.

Paley, N. 2010. Testing electronic detonators to increase sag mill throughput at the Red Dog mine. *Gen. Proc. of the 36th Ann. Conf. on Expl. and Blasting Tech., Feb. 7—10, Orlando,* 2, pp. 49–62.

Potts, G. & Ouchterlony, F. 2006. The capacity of image analysis to measure fragmentation, an evaluation using Split Desktop. Swebrec Report 2005:2, Luleå University of Technology, Department of Civil and Environmental Engineering.

Quezada, D., Muñoz, C., & Arellano, M. 2010. Methodologies for fines control in blasting operations related to porphyry cu supergene materials. *Gen. Proc. of the 36th Ann. Conf. on Expl. and Blasting Tech., Feb. 7–10, Orlando*, pp. 287–292.

Rai, P. & Singh, B. 2004. Investigation of firing patterns on fragmentation in an Indian opencast limestone mine, *Q&M J.*(Feb): 33–37.

Rai, P., Ranjan, A.K. & Singh, B. 2005. Optimizing fragmentation—Investigation of stiffness ratio on fragmentation in moderately hard sandstone strata, *Q&M J.* (Feb.): 33–37.

Rai, P., Ranjan, A.K. & Choudhary, B.S. 2008. Achieving effective fragmentation. *Q&M J.* (Feb): 17–19.

Rai, P., Yang H-S. & Choudhary, B.S. 2012. Formation of slot cut for creating free face in solid limestone bench: A case study. *Powder Technology*, http://dx.doi.org/ 10.1016/j.powtec.2012.05.044.

Raina, A.K., Choudhury, P.B., Ramulu, M., Chakraborty, A.K. & Misra, D.D. 2002. Fragalyst-An indigenous digital image analysis system for grain size measurements in mines. *Journal of the Geological Society of India* 59(6): 561–569.

Raina, A.K., Ramulu, M., Choudhury, P.B., Chakraborty, A.K., Sinha, A., Ramesh-Kumar, B. & Fazal, M. 2010. Productivity improvement in an opencast coal mine in India using digital image analysis technique. In J.A. Sanchidrian (ed.), *Fragmentation by Blasting, Proc. of the 9th Int. Symp. on Rock Fragmentation by Blasting—Fragblast 9, Sept. 13–17, 2009, Granada Spain*, pp. 707–716.

Ramulu, M., Chakraborty, A.K., Raina, A.K., Dudhankar, A.S., Choudhury, P.B. & Jethwa, J.L. 2001. Application of digital image analysis technique for fragmentation assessment and blast optimisation. *Gen. Proc. of the 27th Ann. Conf. on Expl. and Blasting Tech., Jan. 28–31, Orlando*, 1, pp. 273–284.

Sanchidrián, J.A., Segarra, P. & López, L.M. 2006. A practical procedure for the measurement of fragmentation by blasting by image analysis. *Rock Mechanics and Rock Engineering* 39(4): 359–382.

Sanchidrián, J.A., Segarra, P., López, L.M., Ouchterlony, F. & Moser, P. 2010. Evaluation of some distribution functions for describing rock fragmentation data. In J.A. Sanchidrian (ed.), *Fragmentation by Blasting, Proc. of the 9th Int. Symp. on Rock Fragmentation by Blasting—Fragblast 9, Sept. 13–17, 2009, Granada Spain*, pp. 239–248.

Sanchidrián, J.A., Segarra, P., Ouchterlony, F. & López, L.M. 2009. On the accuracy of fragment size measurement by image analysis in combination with some distribution functions. *Rock Mech. Rock Engg.* 42(1): 95–116.

Sarathy, M.O. 2000. Optimum blasting in surface mines—Major issues. *Proc. Sem. on Blasting Objectives and Risk Management, Hyderabad, 13–14 July*, pp. 1–29.

Sarma, S.K. 2000. Mine to Mill blasting to maximize the profitability of mineral industry operations. *Gen. Proc. of the 26th Ann. Conf. on Expl. and Blasting Tech., Feb. 13–16, Anaheim*, pp. 349–359.

Schleifer, J. & Tessier, B. 1996. FRAGSCAN: A tool to measure fragmentation of blasted rock. In J.A. Franklin & T. Katsabanis (eds.), *Measurement of Blast Fragmentation*. Rotterdam, Balkema, pp. 73–78.

Scoble, M.J., Lizotte, Y.C., Singh, A., Hendricks, C. & Mohanty, B. 1991. Fragmentation interaction with underground mining systems and productivity. *Res. Proc. of the 17th Ann. Conf. on Expl. and Blasting Tech., Feb. 3–7, Las Vegas*, pp. 79–90.

Singh, A., Scoble, M. & Lizotte, Y. 2009. Blast fragmentation size analysis techniques and application experience. *Gen. Proc. of the Ann. Conf. on Expl. and Blasting Tech., Feb. 3–7, Las Vegas*, pp. 17–28.

Singh, S.P., VanDoorselaere, D. & Narendrula, R. 2006. A comparison between sieved, predicted, and measured blast fragmentation results. *Gen. Proc. of the 32nd Ann. Conf. on Expl. and Blasting Tech., Jan. 29–Feb. 1, Grapevine*, 2, pp. 263–274.

Thote, N.R. & Singh, D.P. 2003. Blast fragmentation size assessment analysis for production blasts in Indian mining condition. In R. Holmberg (ed.), Explosive & Blasting Technique, *Proc. of the EFEE 2nd World Conf., 10–12 Sept. Prague, Czech Republic*, pp. 459–466.

Thurley, M.J. 2010. Fragmentation size measuring using 3D surface imaging. In J.A. Sanchidrian (ed.), *Fragmentation by Blasting, Proc. of the 9th Intern. Symp. on Rock Fragmentation by Blasting—Fragblast 9, Sept. 13–17, 2009, Granada Spain*, pp. 229–238.

Wang, W.X. & Stephansson, O. 1996. On-line system for monitoring the size distribution of aggregates on a conveyor belt. In J.A. Franklin & T. Katsabanis (eds.), *Measurement of Blast Fragmentation*. Rotterdam, Balkema, pp. 167–174.

Wardrip, L.L., Muller, J.G. & Floyd, J.L. 2002. Improving blast performance with high-density toe loads. *Gen. Proc. of the 28th Ann. Conf. on Expl. and Blasting Tech., Feb. 10–13, Las Vegas*, 2, pp. 167–180.

Workman, L. 2003. The effects of blasting on crushing and grinding efficiency and energy consumption. *Gen. Proc. of the 29th Ann. Conf. on Expl. and Blasting Tech., Feb. 2–5, Nashville*, 1, pp. 131–141.

Yang, H.S. & Rai, P. 2011. Characterization of fragment size vis-à-vis delay timing in quarry blasts, *Powder Technology* 211(1): 120–126.

Measurement and Analysis of Blast Fragmentation – Sanchidrián & Singh (Eds)
© 2013 Taylor & Francis Group, London, ISBN 978-0-415-62140-3

Automated rock fragmentation measurement with close range digital photogrammetry

M.J. Noy
Orica Australia Pty Ltd., Kurri Kurri, NSW, Australia

ABSTRACT: The development of measurement tools for assessing the rock fragmentation induced by blasting has always been constrained by the technology available at the time of development. As computer hardware and software improves, so too do the opportunities to adapt new and evolving technology to the measurement of blast results. One such area of technology that has developed in recent years is desktop photogrammetry. The author began applying photogrammetry (stereo imaging) to the measurement of rock fragmentation on a conveyor in 2005. This technology has recently been adapted for application on an operating face shovel for the fully automated measurement of the rock fragmentation in the muckpile. The use of synchronised stereo cameras for generation of live 3D images has a number of advantages for automated image segmentation algorithms and autonomous image triggering algorithms. This paper discusses the advantage this approach has over standard monocular based fragmentation measurement systems. Also discussed are some of the environmental aspects associated with mounting systems on a shovel. These factors include high vibration and shock, extreme temperatures and weather conditions, and high volumes of dust, all of which are prevalent on an operating shovel in remote and sometimes extreme locations. These environmental concerns have to be addressed to ensure the installed system survives for extended periods without the need for constant maintenance.

1 INTRODUCTION

1.1 *Background*

The expansion of measurement capabilities often depends upon new technology development. Early days in fragmentation measurement (passive visual assessments rather than active sieving of material) involved counting boulders or visually comparing the muck with photographs of piles of rocks with known distributions as in the Compaphoto technique (van Aswegen & Cunningham 1986). The advent and proliferation of personal computers has facilitated an explosion in measurement capability. The transition from analogue to digital data acquisition was underway. Digital cameras began to be widely adopted in the mid to late 1990s. Computer processing power continued to increase rapidly according to Moores Law and image processing algorithms became the standard for processing digital images of rock fragmentation for particle size distribution analysis (Latham et al. 2003).

These algorithms predominantly assessed the two dimensional distribution of the surface fragmentation (Noy 1997). Some systems or photo analysis techniques attempted to construct mass or volume distributions from 1D or 2D measurement using statistical techniques or stereological based "unfolding" functions (King 1982, 1984, Maerz 1996). Surface biases, particle shape variability, particle overlap and image segmentation inaccuracies introduce instabilities in these approaches and limited their usefulness beyond simpler 2D analyses (Outal et al. 2009). In order to circumvent some of these inherent dimensional transformation inaccuracies some researchers attempted to measure the 3D profiles directly. Cheung & Ord (1990) applied light stripes (known as active stereo) to material on a conveyor to depict live 3D profiles. This active stereo, as opposed to passive stereo (dual cameras), was necessary because the available computer power at the time was insufficient to process the passive stereo images in a reasonable time. Attempts were also made to apply active stereo to the muck using Xenon flash strips (Poniewierski et al. 1995). This was aimed at helping with image segmentation and extraction of the 3D distribution but was resolution limited and was quite cumbersome to set up and apply routinely.

At the end of the 1990s digital photogrammetry was beginning to evolve onto the desktop PC

(Konolige 1997). This avenue for 3D profiling of the muck fragmentation (digital terrestrial photogrammetry), did not however, become available until more generic and adaptable software for the PC was developed in the early to mid 2000s (Agrawal et al. 2005). It was the development of the autonomous robotics and surveying industries that was one of the main driving forces behind the evolution of automated software for digital photogrammetry (eg, El-Hakim 1985, Gruen 1992a, b, Konolige 1997, Agrawal et al. 2005, Gong & Yang 2005, Hirschmüller 2005, among others). It then became possible to adapt this software for other measurement purposes, such as for fragmentation measurement.

1.2 *Recent 3D methods used for the measurement of particle size distributions*

Digital photogrammetry offers advantages to image segmentation algorithms when they are applied to fragmentation measurement and provides the ability to determine volumes and potentially derive mass distributions of particle sizes directly. Noy (2006) applied this approach to the measurement of the mass distribution of rock fragmentation on a conveyor. Andersson et al. (2007) successfully applied 3D imaging to the measurement of non-overlapped iron ore pellet sizes with good correspondence to the true sieved size distribution. A 3D laser scanning system has been used to measure the fragmentation distribution from the 3D profile of rock in a Load-Haul-Dump (LHD) bucket (Thurley 2009) and material on a conveyor in a limestone quarry (Thurley 2011). Thurley's (2009) approach, based on Andersson et al. (2007), was to segment the surface rocks in the 3D image and to separate out the overlapped particles from the non-overlapped particles in order to remove the effect of the broadening of the distribution that comes about due to the inclusion of diminished overlapped particles. This effect is particularly relevant for iron ore pellets where the size distribution is generally very narrow and the particle shapes convex, hence second layer particles are almost always diminished by being partially obscured by top layer particles. However, this is not as dominant for the broader distributions prevalent in rock piles, because of the opportunity for smaller rocks to fit in between larger rocks on the surface.

There are some difficult challenges when applying a photogrammetric system developed for a conveyor to the automated measurement of the fragmentation in the muck pile from an operating shovel. This paper highlights some of these challenges and discusses the advantages of such a system over the simpler single camera systems.

2 CHALLENGES IN FULLY AUTOMATING FRAGMENTATION MEASUREMENT

2.1 *Image segmentation challenges*

Measurement of rock fragmentation by analysing digital photographs is a difficult task to fully automate. Segmentation algorithms need to be sufficiently robust to handle images of rocks in varying lighting conditions, surface reflectivity, texture and presentation. Figure 1 shows examples of different sample presentations that highlight some of these challenges.

The difficulty for 2D algorithms is that the light dependent image texture is the only information available to identify particles and segment then correctly. In Figure 1a the texture of fines is very similar to the texture on the tops of the larger rocks. That problem can make computer based decisions on segmentation less reliable. When dealing with material on conveyors there is also the problem of having to differentiate the belt (Fig. 1b) and not confuse the belt texture with particles or fines. When on an operating shovel the situation becomes even more difficult (see Fig. 1c, d). A fully automated system needs to contend with varying lighting conditions, occasional interrupted viewing, dust, rain or snow in the field of view, varying distances to the muck both within and between images and possible motion blur in low light conditions. Under these conditions, fully automated capture and analysis from a shovel becomes a very difficult task, especially when light dependent

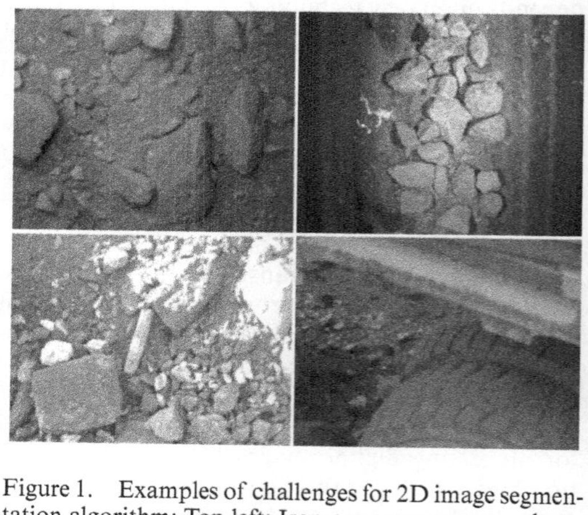

Figure 1. Examples of challenges for 2D image segmentation algorithm: Top left: Iron ore on a conveyor showing similar texture between fines and the surface of large rocks; Top right: rocks on a conveyor where the conveyor belt is exposed in the image; Bottom left: variable lighting across an image at the muck face taken from an operating shovel; and Bottom right: unplanned interruption to the view of the fragmentation in the muck.

image texture is the only information available for making segmentation decisions.

2.2 *Image scaling challenges*

The varying distance to the muck rill requires that the range be measured so images can be appropriately scaled. Figure 2 shows the position of a camera system on a P&H 4100 electric rope shovel and a chart showing the range to the muck rill during bucket engagement in the muck. For fixed lens conditions the area mapped by a pixel in the captured image(s) is a function of range. For a 25 mm focal length lens and a sensor size of 16.9 mm (1360 × 1024 pixels), the pixel size mapped in the image varies from 3.1 mm × 3.1 mm to 3.8 mm × 3.8 mm for capture ranges between 12 m and 14.5 m. This translates to approximately an overall 20% error in the particle sizing if range is not taken into account.

If triggering is independent of the shovel operating position the muck range during shovel operation can vary from between 11.5 m and 15 m and this relates to an approximate error of 31% in the mapped pixel size. For example, a particle of 1 m diameter at a 15 m range would present as a particle with an apparent diameter of 1.31 m at an 11.5 m range for a fixed lens, single set calibrated pixel size.

Perspective distortion in 2D images is a problem for two reasons: (1) a single scale for the image pixels is no longer valid across the entire image; and (2) the rock protrusions from the surface of the muck obscure particles behind affecting the visible projected particle size distribution apparent in the image. The latter cannot be corrected for by varying the scale across the image. Unless 3D mapping of the rill surface is done within an image it is often difficult to assess whether perspective distortion is even present. Figure 3 shows the within-image range variation measured from 3D images captured during bucket engagement (P&H 4100 shovel) in the muck (sampled from 3 half hour sessions over 3 days). Given that the variation in range within an image can be as large as 3.5 m (~31% error), the validity of using an image with such a range variation is questionable, not just because of the scaling variability across the image but also because of the angle dependence in the apparent distribution arising from the increased incidence of particle overlap and occlusion of smaller/lower profile particles on the surface. It is also instructive to note that the smallest variation in range across an image (when bucket is engaged in the muck) is 1 m. This range variation relates to the fragments on the surface of a planar and perpendicular rill. If this was due to a sloping background and not purely the rock fragments then this would equate to around a maximum 8% error in the scaling of fragments between the extremes of the measured

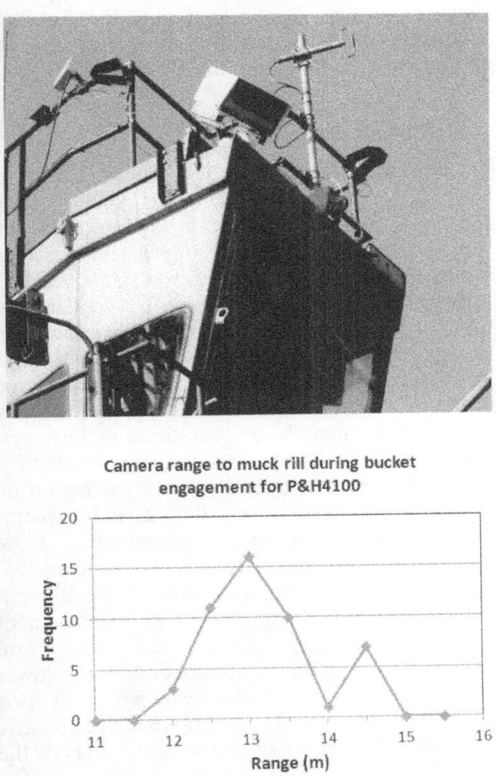

Figure 2. Range from the camera on the top of the cabin of a P&H 4100 electric rope shovel to the muck rill during bucket engagement. The distance is measured to the muck rill just to the right of the bucket.

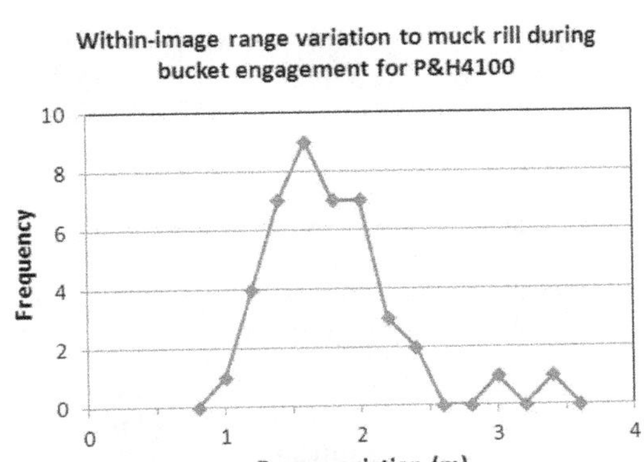

Figure 3. The graph shows the maximum variation in range within images sampled during bucket engagement in the muck.

distance in the image. There is a maximum 13% error in the scaling of fragments at the extremes of the background range variation within images with variation of 1.8 m, the average depicted in the 3 sampled half hour sessions shown in Figure 3. However, this calculation does not include the error due to increased particle overlap apparent in oblique views of the rill surface.

Lens distortion is another issue with single camera systems. Correction for lens distortion is rarely used in cameras capturing images of the fragmentation in a muck pile. This error can be a small depending on the quality of the lens and the focal length used. However, if wide viewing angles are used there can be a strong distortion both due to the lens and perspective at the periphery of an image. This area should be avoided when quantifying the fragmentation distribution from such images.

Given these issues, it is also apparent that the triggering of images from the shovel bucket position and assuming an average range, is likely to incur significant particle scaling errors. Single point range measurement during the image trigger will help to minimize these inter-image scaling errors. However, in order to account for intra-image range variations, the trigger of images should depend on the full 3D range profile so that steep profiles across the area being viewed are avoided. Furthermore, image segmentation based solely on image texture, which is subject to both rock texture and lighting conditions and possibly subject to foreign objects like vehicles in the field of view, may be unreliable because of the limitations in any algorithm attempting to automatically extract the critical content based on textural information. These problems can be overcome if real-time 3D triggering and analysis is employed.

3 DEVELOPMENT OF AN AUTOMATED FRAGMENTATION MEASUREMENT SYSTEM USING STEREO IMAGING

3.1 *Conveyor based system*

Computer hardware and software development in recent years have underpinned the delivery of desktop photogrammetry to the broader community. The adaptation of this capability to the mapping of surface profiles of rock piles can deliver the additional information required for a more reliable triggering and segmentation of images and consequently the quantification of the rock fragmentation size distribution.

The application of photogrammetry to fragmentation measurement has also highlighted a requirement for camera calibration. The process of setting up a stereo vision system for gathering 3D profiles of rocks requires the calibration of the camera pair. This calibration process corrects for camera and lens distortions and hence directly deals with one aspect of image based measurement of fragmentation that is often overlooked in practice with single camera systems.

Application of passive digital terrestrial photogrammetry to measurement of rock particles on a conveyor was reported by Noy (2006). This application of photogrammetry allows the determination of the particle volume distribution because the system algorithm can remember or estimate the floor of the conveyor belt (ie, the 3D profile of the belt without rocks) in determining the volume of material above the belt. Therefore, given that the particle segmentation derives a volume as a function of particle size, the mass distribution can easily be determined if particle density is known across the distribution. In practice this density adjustment can be done through calibration of multiple samples presented to the system. Figure 4 shows a sample of a series of belt cut comparisons with automated analyses taken over a 7 month period below a Conveyor Vision System (CVS) installed in a quarry in Europe. Figure 5 shows some typical 3D images generated from this CVS.

3.2 *Shovel based systems*

The application of the CVS to the measurement of the fragmentation in the muck pile from an active face shovel has required some modifications to the system. In the first instance, the placement of a calibrated stereo imaging device on an operating shovel requires the system to be effectively isolated from the vibration of the shovel. Vibration both damages the components and potentially disrupts the correspondence between the calibrated stereo images through differential motion of the cameras. Other environmental considerations include system temperature control. Most components of a vision system have a limited operating temperature range and as a result require the internal temperature of the system to be maintained within these limits. Moisture ingress is also a consideration. Figure 6 shows the latest version of the Binocular Vision System (BVS) developed for placement on an operating face shovel. Clearly visible are the vibration isolation mounts between the lower U-bracket and the upper H-bracket which the BVS enclosure clips into. These isolators are roughly matched to the weight and typical frequency of the load and are orientated for absorption along the major axis of vibration on a shovel.

Also evident is the presence of a large heat sink relating to the Peltier temperature control system. This system operates to keep the internal tempera-

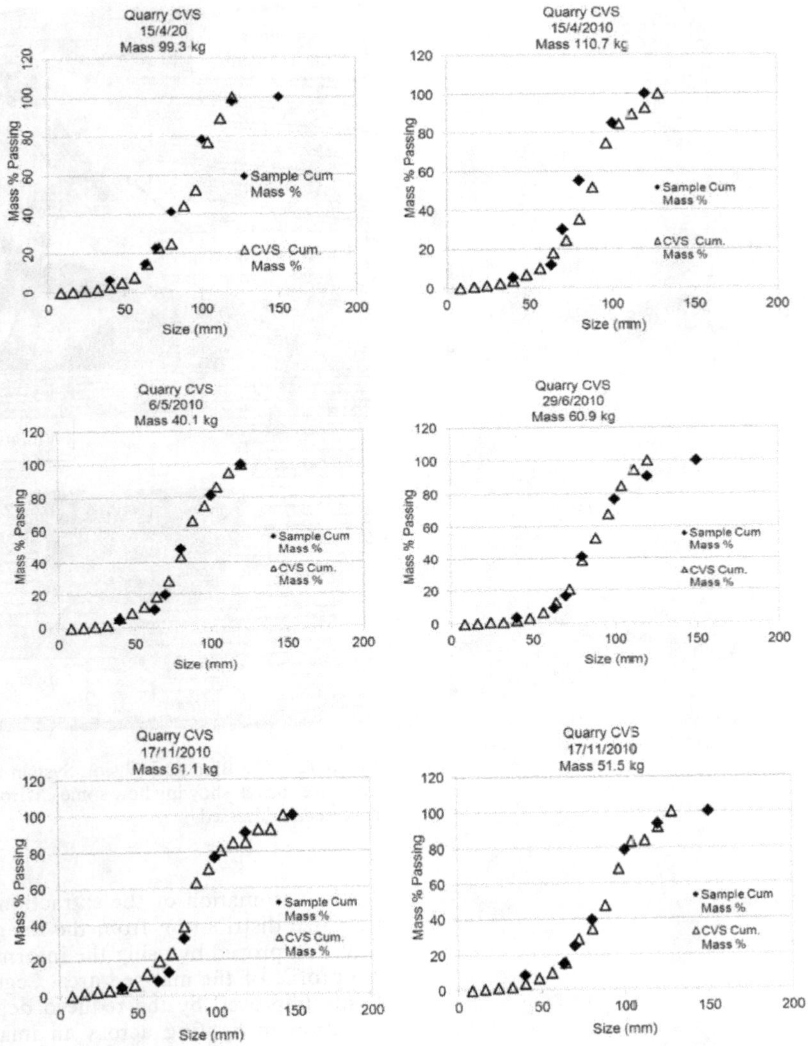

Figure 4. A series of mass distributions measured from sampled belt cuts (solid diamonds) taken from under a CVS (Conveyor Vision System) compared with the automated CVS measurements (open triangles) of the exact sample material before being removed.

ture within the limits of the most temperature limited component in the system: the cameras. The hood also assists in preventing direct sunlight from heating the enclosure. A hydrophobic vent is also employed to prevent moisture ingress.

As mentioned in the previous section, triggering the system to capture images requires knowledge of the distance to the muck rill. Utilising the 3D range profile for triggering the camera allows a 3D system to ensure the plane of the muck in the image captured is relatively perpendicular to the camera axis, thereby minimising the scaling and oblique angle distortion errors associated with the variation in range within an image. Triggering based on range information gathered from the live 3D profile generated also makes the system completely autonomous, not requiring any information about bucket position from the shovel, so that it is easily applied to any make of shovel without requiring any specific shovel input for triggering purposes. Range information used in triggering also ensures that the system corrects for range variation between images, thus minimizing the error in scaling to less than the within-image range error

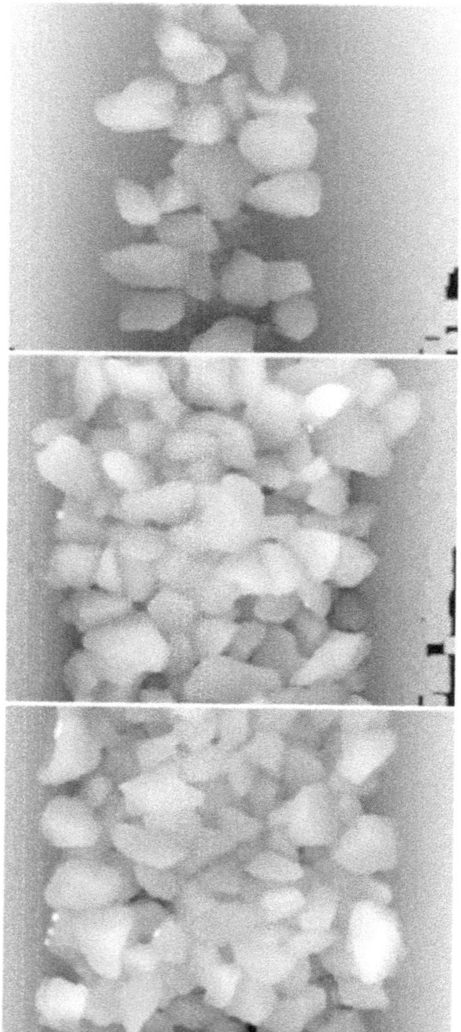

Figure 5. Example of 3D images generated from material passing below the CVS on a conveyor travelling at around 3 m/s. The higher the profile the lighter the pixel. Generally a set range measured by the CVS is mapped into the 8-bit grayscale range of the image.

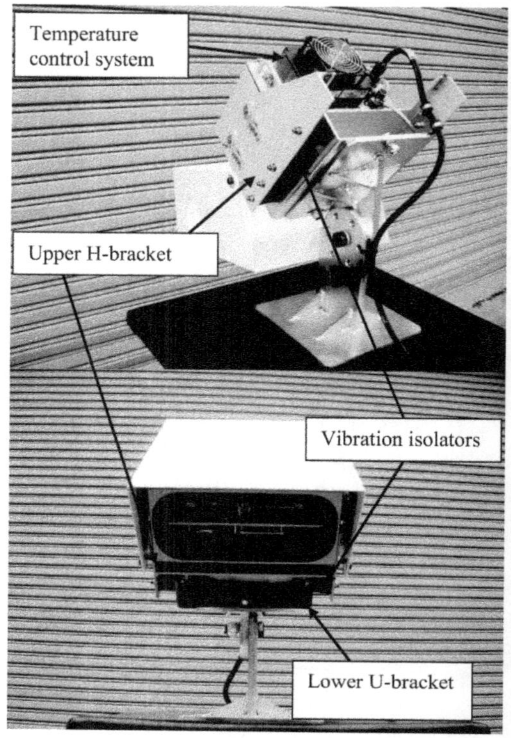

Figure 6. The Binocular Vision System for installation on a face shovel, showing how some environmental issues have been addressed.

(minimized by constraining the trigger acquisition of images to relatively perpendicular planar muck views). This amounts to a system that can capture images of the muck fragmentation with minimal error associated with image lens distortion, perspective distortion and scaling of particles. The system should therefore be more sensitive to changes in the fragmentation size distribution than alternate systems that do not account for these factors.

The automation of the extraction of the particle size distribution from the images captured is also improved by using the information in the 3D profile of the mapped area. Segmentation is firstly improved by the reduced dependence on variation in lighting across an image which is often encountered in images captured from an operating shovel. The construction of a "3D image" (i.e., mapping range to muck into the 8-bit range of an image) depends on the distinguishing of corresponding areas between the stereo image pair. Provided that lighting is sufficient to identify rock texture and not too bright so as to saturate the image, the photogrammetric algorithm can generate the 3D profile for the intersecting image regions on the muck rill. This profile is then clear of raw image textural variations and hence can be used to extract the fragmentation size distribution without lighting or rock texture noise to interfere with the segmentation. Figure 7 shows a typical image collected from an operating shovel with the shadow of the cabin in view and the corresponding 3D image generated. Clearly the automated photogrammetric analysis

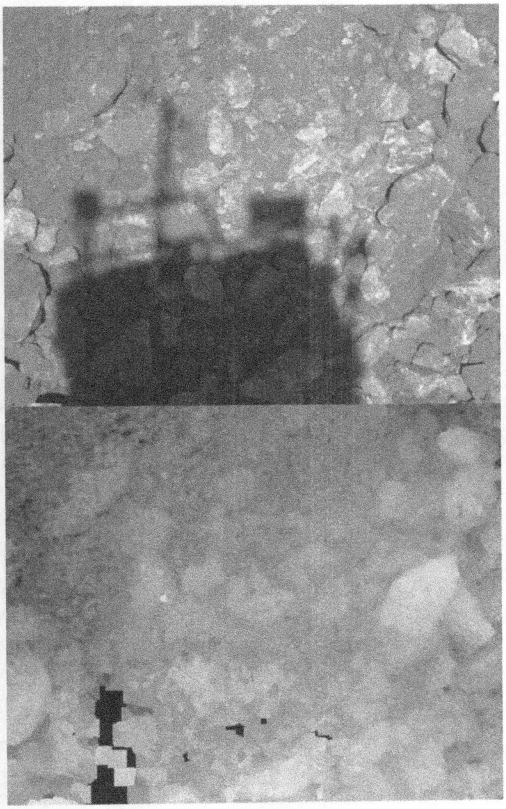

Figure 7. Light related texture (shadowing) and the corresponding 3D image showing a clearer view of the information of interest. Minor aberrations are still present and indicate a localised loss of textural information between the stereo image pair.

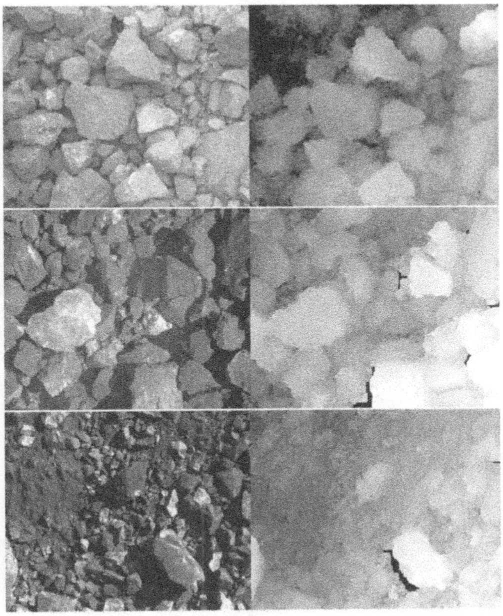

Figure 8. Typical images captured from an operating shovel and the corresponding 3D images generated.

is quite robust given the lighting variations across the image.

Any loss of textural information is evident as holes in the 3D image. Small holes can be addressed to some extent by interpolation from surrounding information; however, larger holes are more difficult to interpolate across and are left to indicate loss of information in that region.

Figure 8 shows some typical images captured from a BVS installed on a shovel and the corresponding 3D images that were generated. The dark patch in the top left hand corner of the first 3D image indicates that the range is outside the allocated range mapped into the image. Saturation in the second 3D image also indicates that the range is closer than the allocated range mapped into the image. These 3D images show that the variation in range across an image may not be obvious in the 2D raw image as is the case with a single camera system acquisition and that multipoint range measurement is required to ensure areas sampled are relatively planar and normal to the camera to minimise calibration errors. The last pair of images in Figure 8 show the limitation in the 3D approach to fragmentation measurement. This relates to the depth perception of the current system at ranges beyond 11 m. Synchronised stereo imaging provides depth information by looking for disparities between the image pairs. This is the displacement of the same objects on the two imaging sensor due to the separation (and orientation) of the cameras (roughly referred to as the base or baseline) and other camera specific settings (e.g., lens focal length, sensor size and pixel resolution). For parallel alignment of cameras: the larger the disparity the closer the object. Zero disparity would relate to objects at infinity. However, given the quantization of sensors (discrete size of pixels), this zero disparity approximates to distances closer than infinity. Figure 9 shows a comparison between an automated analysis and a manual analysis of a few randomly selected images acquired by a BVS system on an operating hydraulic shovel. The BVS auto-analysis algorithm currently accumulates more fines but converges on the upper size fraction of the distribution allowing a good aggregated P80 to be determined. The limited depth perception of the system at shovel to muck distances currently

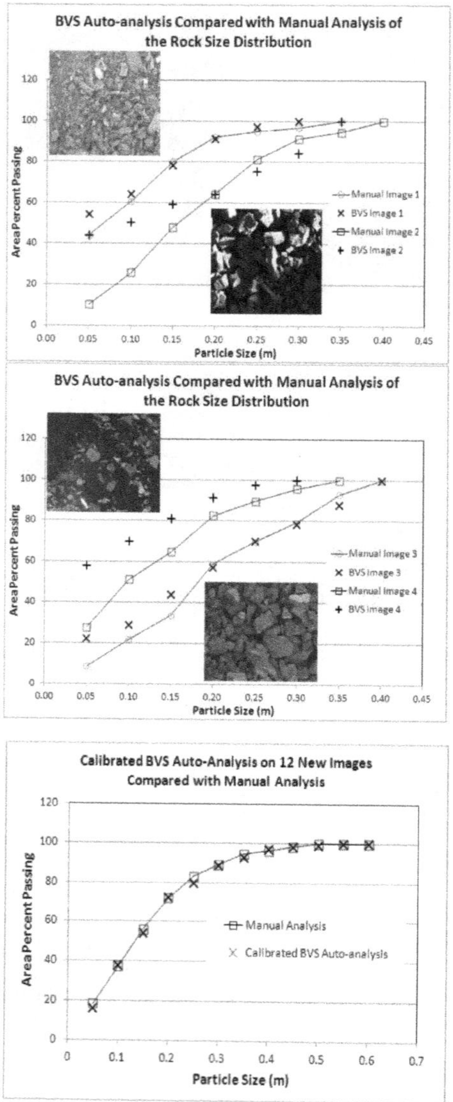

Figure 9. Comparison of a BVS analysis of four images with manual analysis using PowerSieve™ software. Image 1 pixel size = 2.06 mm; Image 2 pixel size = 1.96 mm; Image 3 pixel size = 1.89 mm; Image 4 pixel size = 2.14 mm. The bottom graph shows the calibrated BVS algorithm performance on 12 new BVS images compared with a manual analysis of the same images.

limits the accurate determination of fines fractions on the muck surface. Calibration of the BVS algorithm is then achieved using the distribution from the more accurate manual analysis of these four images. The bottom graph in Figure 9 shows the result of the calibrated BVS auto-analysis on 12 new images and compares this with a subsequent manual analysis of the same images.

Note that the current BVS system has a range resolution of 4 cm at 13 m. This limits the ability to identify small bumps in the 3D profile (note that these particles are currently gathered as fines). At this stage of development, calibration of the analysis is required to achieve a more accurate measurement of the full distribution. However, improving the range resolution of the BVS at 13 m is still desirable to ultimately eliminate the need for calibration. This improvement can be achieved by modifying the current stereo camera configuration, by improving the existing camera resolution, computer processing power and algorithm sophistication. Increasing the baseline between the cameras will improve range resolution but such a system is more likely to be subject to differential motion between the stereo camera pair due to shovel vibration. This can interrupt the correspondence between the cameras and result in the loss or degradation of the 3D information generated by the automated photogrammetric algorithm. Hence the baseline length will be limited by how stiff the system can be made. The higher the camera resolution, the better the range resolution, but this comes with more significant demands on computer processing power. Ultimately, a combination of improvements in the camera baseline length, camera resolution, including camera thermal and vibration robustness, and computer power will help improve the depth perception of the system. These next steps in development will eventually be met by the onward advances in technology more generally.

4 CONCLUSIONS

Photogrammetry offers tangible benefits to the measurement of rock fragment size distributions on conveyors and in muck piles. Full automation of fragmentation measurement can take advantage of the additional information offered by rill profiles generated by stereo imaging systems. For example, image capture can be constrained to appropriate rill presentations such that the within-image range variation is not too large so that the scale remains relevant across the entire image. Image segmentation algorithms can benefit from the extra dimensional information obtained above that of 2D systems, thereby not having to rely solely on image texture (area illumination, rill surface texture, etc.) alone for identifying particles. The depth perception of the BVS is limited for large shovel to muck distances, but this can be compensated for with appropriate measurement calibration. The camera systems that are installed on face shovels require protection from relatively harsh environmental conditions. These issues can be addressed through sensible engineering.

ACKNOWLEDGEMENTS

Thanks to Nicholas Graen and Philip Buttgereit, Orica Germany, for collecting and measuring belt cuts below the CVS in Europe. Thanks also to Startek Technology Pty Ltd for their contribution to the development of the Orica Vision Systems.

REFERENCES

Agrawal, M., Konolige, K. & Iocchi, L. 2005. Real-time detection of independent motion using stereo. *Proc. IEEE Workshop on Motion and Video Computing, Breckenridge, 5–7 January, 1*, pp. 207–214.

Andersson, T., Thurley, M. & Marklund, O. 2007. Visibility classification of pellets in piles for sizing without overlapped particle error. *9th Biennial Conference of the Australian Pattern Recognition Society on Digital Image Computing Techniques and Applications (DICTA), Glenelg, 3–5 December*, pp. 508–514.

Cheung, C.C. & Ord, A. 1990. An on line fragment size analyser using image processing techniques. *FRAGBLAST 3—Third International Symposium on Rock Fragmentation by Blasting, Brisbane, 26–31 August*, pp. 233–238.

El-Hakim, S.F. 1985. A photogrammetric vision system for robots. *Photogrammetric Engineering and Remote Sensing* 51(5): 545–552.

Gong, M. & Yang, Y-H. 2005. Near real-time reliable stereo matching using programmable graphics hardware. *Proceedings of Computer Vision and Pattern Recognition (CVPR)* 1: 924–931.

Gruen, A. 1992a. Tracking moving objects with digital photogrammetric systems. *Photogrammetric Record* 14(80): 171–185.

Gruen, A. 1992b. Recent advances of photogrammetry in robot vision. *ISPRS Journal of Photogrammetry and Remote Sensing* 47: 307–323.

Hirschmüller, H. 2005. Accurate and efficient stereo processing by semi-global matching and mutual information. *Proc. CVPR* 2: 807–814.

King, R.P. 1982. Determination of the distribution of size of irregularly shaped particles from measurements on sections or projected areas. *Powder Technology* 32: 87–100.

King, R.P. 1984. Measurement of particle size distribution by image analyser. *Powder Technology* 39: 279–289.

Konolige, K. 1997. Small vision systems: hardware and implementation. *Eighth International Symposium on Robotics Research*, pp. 111–116.

Latham, J-P., Kemeny, J., Maerz, N., Noy, M., Schleifer, J., & Tose, S. 2003. A blind comparison between results of four image analysis systems using a photo-library of piles of sieved fragment. *International Journal for Blasting and Fragmentation* 7(2): 105–132.

Maerz, N.H. 1996. Reconstructing 3-D block size distributions from 2-D measurements on sections. In J.A. Franklin & T. Katsabanis (eds), *Measurement of Blast Fragmentation, Proc. Fragblast-5 Workshop on Measurement of Blast Fragmentation, Montreal, 23–24 August*, pp. 39–43.

Noy, M.J. 1997. 2D versus 3D fragmentation analysis: preliminary findings. *Proc. 13th Annual Symposium on Explosives and Blasting Research, 2–5 February, Las Vegas*, pp. 181–190.

Noy, M.J. 2006. The latest in on-line fragmentation measurement—stereo imaging over a conveyor. *FRAGBLAST 8—Proc. 8th International Symposium on Rock Fragmentation by Blasting, 7–11 May, Santiago*, pp. 61–66.

Outal, S., Schleifer, J. & Pirard, E. 2009. Evaluating a calibration method for the estimation of fragmented rock 3D-size-distribution out of 2D images. *FRAGBLAST 9—Proc. 9th International Symposium on Rock Fragmentation by Blasting, Granada, 13–17 September*, pp. 221–228.

Poniewierski, J., Cheung, L.C.C. & Maconochie, A.P. 1995. SIROFRAG—A new technique for post-blast rock fragmentation size distribution measurement. *Explo Conference, Brisbane, 4–7 September*, pp. 263–267.

Thurley, M.J. 2009. Fragmentation size measurement in LHD buckets using 3D surface imaging. *FRAGBLAST 9—Proc. 9th International Symposium on Rock Fragmentation by Blasting, Granada, 13–17 September*, pp. 229–237.

Thurley, M.J. 2011. Automated online measurement of limestone particle size distributions using 3D range data. *Journal of Process Control* 21(2): 254–262.

van Aswegen, H. & Cunningham, C.V.B. 1986. The estimation of fragmentation in blast muckpiles by means of standard photographs. *J.S.A.F. IMM.* 9(2): 144–148.

Measurement and Analysis of Blast Fragmentation – Sanchidrián & Singh (Eds)
 ISBN 978-0-415-62140-3

Automated, on-line, calibration-free, particle size measurement using 3D profile data

M.J. Thurley
Luleå University of Technology, Luleå, Sweden

ABSTRACT: Optimisation and control of blasting, comminution and agglomeration processes is a complex task with large potential for gains in energy efficiency and productivity in the mining and aggregates industries. In order to realise these benefits, fully automated, non-contact, on-line particle size measurement technology is required to provide the necessary feedback. Results from two installations are presented for measurement of crushed rock on conveyor belt from a variety of sieved products between 0 and 100 mm and material from a primary crusher up to 250 mm. Detailed sieve-size distribution results are presented with results calculated directly from the 3D surface profile data of the conveyor. No calibration against sieve samples has been performed clearly demonstrating the capability of the system to operate in a plug-and-play low maintenance setup. The technology measures 3D surface profile data which is used to delineate individual particles, identify non-overlapped particles, identify areas-of-fines, and produce a sieve-size distribution.

1 INTRODUCTION

1.1 *Overview*

In the mining and aggregate industries a great many processes affect, and are affected by particle size, including blasting, comminution, and agglomeration. As a result significant effort goes into measuring or estimating the size distribution of particulate material in order to evaluate and optimise production in terms of production rate and production costs (energy, material and equipment costs).

Mine and quarry operators want to measure the particle sizing results of all of these activities but sieving/screening is an imperfect assessment tool due to slow feedback, and inconsistent measurement due to operator fatigue, or variations in technique.

As a result there is an opportunity for on-line, non-contact, fully automated machine vision systems for measurement of particle size to facilitate evaluation and optimisation of mining and particle processes.

1.2 *Source of error*

There are however, a number of sources of error relevant to techniques that measure only what is visible on the surface of a pile and it is necessary to consider these errors in order to ensure a measurement system can be stable, reliable, and trend in the right direction.

Particle delineation error refers to the inaccuracies of determining the correct delineation of all the individual particles in the measured surface (whether by an automatic computer program or manually). Significant error will render the measured surface largely meaningless as the particle delineation will bear little resemblance to the reality of what is on the surface of the pile. This error is significantly related to the type of surface measurement technique applied, whether photographic 2D, or 3D surface measurement. Given the impact of this error, evaluation of particle delineation is highly important. This error has been evaluated for the key particle delineation methods used here with 3D surface profile data by Thurley and Ng (2005).

Sub-resolution particle error, relates to the inability of an imaging system to see fine particles below the resolution of the sensor. For example, to detect an individual particle that is only represented by a few pixels or less in an image. These sub-resolution particles tend to be grouped into large regions and mis-sized as large rocks.

Segregation and grouping error, more generally known as the Brazil nut effect (Rosato et al. 1987), describes the tendency of the pile to separate into groups of similarly sized particles. It is caused by vibration or motion (for example as rocks are transported by truck or conveyor) with large particles being moved to the surface. It is advisable to measure at a point early on the conveyor before the material has been subjected to excessive vibration and segregation.

Overlapped particle error describes the fact that many particles are overlapped (refer to Fig. 1) and are only partially visible. A bias towards the

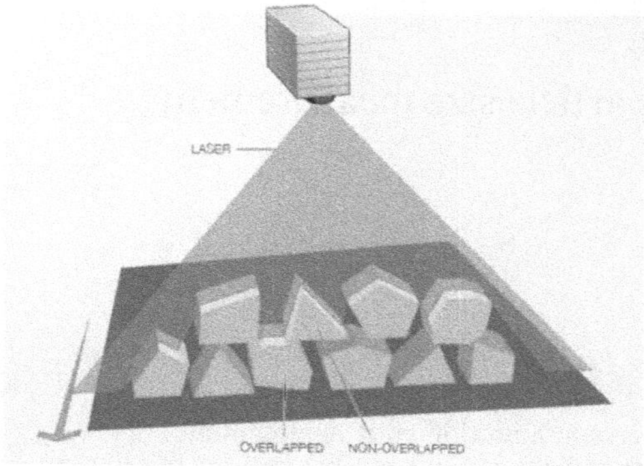

Figure 1. Illustration of overlapped and non-overlapped particles on conveyor passing through a sheet of laser light.

smaller size classes results if overlapped are treated as small non-overlapped and sized using only their visible profile. This error can be overcome in piles of particulate material using classification algorithms based on 3D range data (Thurley & Ng 2008) successfully providing 82% classification accuracy on hold-out data (Andersson & Thurley 2008).

Capturing error describes the varying probability based on size that a particle will appear on the surface of the pile. In simple terms, the larger a particle is, the more likely one is to be able to see some part of it on the surface. For example, if a single particle is as large as the height of the pile of material, then it will always be visible, whereas a very fine particle is almost certainly not visible. Thurley (2002) has explored capturing error in laboratory rock piles but it remains a source of error in this application.

Profile error describes the fact that only one side (a profile) of a non-overlapped particle can be seen making it difficult to estimate the particles size. In the presented research the partial 3D surface profile of each non-overlapped particle and depth of the bed of material is used to estimate a 3D ellipsoidal volume of the particle.

Sample delimitation and extraction is an error relevant to all methods sampling from conveyor. Overcoming this error requires the correct delimitation of a belt section using two parallel transverse cuts across the belt, and correct extraction of particles where only particles whose centre-of-gravity is inside the delimited region are part of the sample. Refer to Pitard (1993) for a more thorough description.

Weight estimation error results from the fundamental difference between non-contact measurement and physical measurement. Size measurement using imaging identifies how many particles are observed, but manual sieving measures the weight of particles in each size class. Therefore it is necessary to have a method of mapping from numbers of particles to weight of particles in order to provide a measurement of size that industry understands and can use. The presented research uses volumetric estimation of each non-overlapped particle and assumes constant density within a sample to estimate a weight of each particle. Weight of fines is estimated based on the bulk volume of the observed areas-of-fines.

1.3 *Literature review*

Particle size measurement using vision has been the subject of research and development since the 1980's (Carlsson & Nyberg 1983, Ord 1988) with a legacy of predominantly photographic based systems with widely varying degrees of success and no general solution available on the market.

Photographic based 2D imaging systems are subject to significant *particle delineation error* due to uneven lighting conditions, excessive shadowing, and colour and texture variation in the material. Furthermore, photographic systems have no direct measure of scale, suffer from perspective distortion, lack the capability to distinguish between overlapped and non-overlapped particles, and do not demonstrate the ability to automatically detect visible fines in a realistic way. As a result photographic 2D systems typically require manual editing of the particle delineation to provide a reasonable estimation.

In 1996 the Fragblast 5 conference held a workshop on the measurement of blast fragmentation with the stated aim "to review the state-of-the-art in image processing as applied to measurements of rock fragmentation" (Franklin & Katsabanis 1996). Prior to Fragblast 10, this was the most thorough review of systems (photographic 2D) for this purpose. The keynote address from this workshop by Cunningham (1996) is particularly critical in its review of image analysis systems for rock fragmentation and in highlighting common conclusions. Cunningham (1996) noted "that almost every such paper includes a statement such as *the system is practical for suitable use … although further work is required*". Furthermore, "graphs of system output versus sieving results are described as providing *good* or *reasonable* correlation" but that "it is often difficult to concur with the sentiments expressed".

In their review of the Split commercial photographic based 2D system Potts & Ouchterlony (2005) report that for their application the system erroneously assumes the resultant size distribution is uni-modal and they conclude by expressing strong reservations saying 2D "imaging has a certain but limited usefulness when measuring the fragment size distribution in a muckpile or from a belt in an accurate way. It could probably detect rough tendencies in fragmentation variations, if the lighting conditions do not vary too much, and if cover glasses for camera lenses are kept clean".

Comparisons have been published between photographic methods and sieving, specifically Wang & Stephansson (1996) and Fernlund (1998). Wang & Stephansson (1996) performed image analysis on overlapping fragments and reported "a systematic error compared to sieving analysis". This error is in part due to the inability of photographic 2D image analysis and particle delineation to provide the information necessary to distinguish between entirely (non-overlapped) and partially visible (overlapped) rocks.

There are a number of publications relating to 3D size measurement including Noy (crushed rock, 2006), Frydendal & Jones (sugar beets, 1998), Kim et al. (river rock, 2003), Lee et al. (2005), and Thurley (2002, 2009, 2011). However, Frydendal and Jones (1998) and the presenting author (Thurley & Ng 2008) are the only publications to remove the bias resulting from overlapped particles. For conveyor belt applications some publications recommend installing a mechanical vibration feeder (Kim et al. 2003, Lee et al. 2005) to separate rocks and prevent particle overlap. If there exists the possibility to install a vibration feeder at full-scale in the process it would simplify image analysis, and would likely improve results but with the added cost and maintenance of the additional equipment.

The focus of this research is on systems for online measurement of the particle bed that are non-contact, that is they do not require any additional material handling. If non-contact measurement is required then one must consider overlapped particle error and account for overlapped and non-overlapped particles. Furthermore, in some circumstances, such as examination of rocks in in-production excavators (Thurley 2009), there is no other option than to account for overlapped and non-overlapped particles.

1.4 *Measurement system overview*

This research uses an industrial measurement system on conveyor belt based on laser triangulation (a projected laser line and camera at an offset angle) collecting highly accurate 3D profiles of the laser line approximately 3000 Hz. This high speed ensures a high density of 3D point data with a spatial resolution between consecutive points in the direction of the belt of approximately 1 mm for a conveyor belt is running at 3 m/s. One measurement system is installed at a limestone quarry on the conveyor belt used for ship loading and measures the material on the belt during loading every 30 seconds. This allows measurement of a variety of screened and stockpiled products with a narrow range of sizes and it is against these products the measurement system performance is evaluated. Figure 2 shows an image of the installed laser and camera mounted above the conveyor belt as illustrated in Figure 1.

Using the same laser triangulation technology it is also possible to collect measurement points at 0.1 mm spatial resolution and 0.005 mm depth resolution as we have done on steel slabs. Using such a setup it should be possible to confidently measure particles with a lower size limit of 1 mm and be able to detect overlapped and non-overlapped particles on conveyor belt. An equivalent reduction in measurement speed (one tenth conveyor belt speed) is necessary such as 0.3 m/s but it is theoretically possible to improve this using higher power lasers (beyond class 3B).

The presented analysis algorithms are not particularly dependent on the laser triangulation measurement technology. Any other technique for capturing 3D surface data of a particle pile, such as stereo photogrammetry, laser scanners, or time-of-flight 3D cameras could also be used.

The computational speed of the analysis process is approximately 20 seconds on a 2.7 GHz Intel i7 CPU processing a data set of approximately 1 million 3D points (1.5 m long, 800 mm wide section of the belt). Furthermore, advances in both algorithmic efficiency and hardware are both available to improve computational time as the image processing operations are readily parallelisable. Computation time could be reduced down to the order of few seconds or less for rapid automatic control applications. Once the analysis

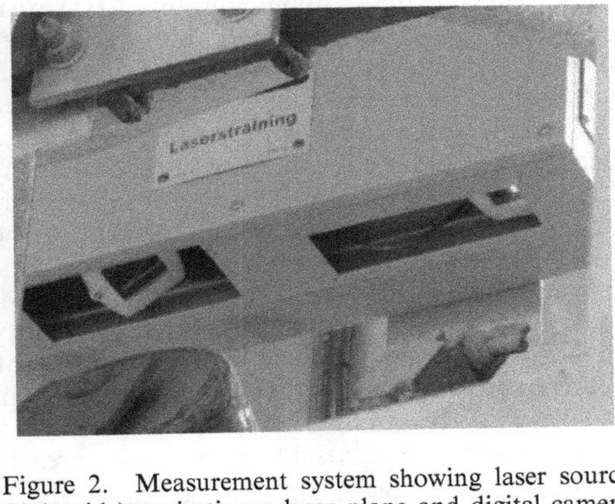

Figure 2. Measurement system showing laser source (right side) projecting a laser plane and digital camera (left side) observing the interaction of the laser and the particles on the conveyor.

can be performed in a under a second then it will to become possible to measure and analyse the entire surface of the material passing under the measurement system.

It is necessary to consider the largest particle size that should be measured and how many of these particles need to be measured to get a statistically valid measurement size. Given the circumstances in this application we currently use a 1.5 m long, 800 mm wide conveyor belt sample is sufficient given the material top size of approximately 90 to 100 mm.

1.5 *Research background*

The presented research builds upon a series of achievements and research developed on both laboratory rock piles and industrial application.

An industrial measurement system on conveyor belt for iron ore pellets (Thurley & Andersson 2007) has been developed using the same laser triangulation measurement technology. The high speed camera system ensures we have a high density of 3D point data at a spacing between consecutive points in the direction of the belt of approximately 0.5 mm. This high data density has at least two advantages. Firstly it allows us to detect small sliver regions or crescent-like regions of overlapped particles and ensure that they are not merged into other regions. Secondly, it ensures high resolution when it comes to measuring the size of each iron ore pellet allowing a size distribution with very fine spacing of 5, 8, 9, 10, 11, 12.5, 13, 14, and 16+ mm size classes.

One of the key criteria for particle size measurement is therefore high data density as it defines the capacity to detect small overlapped particles, the lower limit on individual particles that can be reliably detected, and the resolution of size classes detectable.

In addition a demonstration project for size measurement of rocks in underground LHD excavator buckets (Thurley 2009) has been performed. A 3D vision system based on laser scanners was installed on the tunnel roof in a production area of an underground iron ore mine with 3D surface data of the bucket contents being collected as the LHD unit passes beneath. The project successfully demonstrated fragmentation measurement of the rocks in the bucket, identifying overlapped rocks, non-overlapped rocks, and estimation of the sieve-size of the visible rocks. The potential to identify areas-of-fines and prevent misclassification of such regions as large rocks was shown, and used in the presented research.

This research presents advances in three areas;

1. development of an image analysis strategy that works across a broad range of material scales from fine particles (0.2 mm), up to (but not limited to) 250 mm. A summary of these results is presented but the details of this process are outside the scope of this paper. Background research has been previously published by Thurley (2011) and Thurley & Ng (2005).
2. validation of the capacity to detect areas-of-fines using a variety of test data, and
3. development of a volumetric based particle sizing method that individually sizes each non-overlapped particle, detects areas-of-fines, and incorporates this information into a sieve-size-distribution (by volume). Assuming constant density of the material, this provides an estimation of the sieve-size-distribution by weight.

2 METHOD & RESULTS

2.1 *Data collection*

In the installation at the limestone quarry the majority of particles are above 10 mm and the system was originally installed to only measure particles at 10 mm or larger. Therefore the system collected 3D data points at a spatial resolution of 1 mm to ensure sufficient data resolution to identify these particles. This spatial resolution is maintained for the second measurement system located after the crusher so we can easily compare results between the two systems.

A number of different products (size ranges of material) were measured in order to evaluate the system at various scales. Particularly important was to test the system with material below the measurement limit of the system. As the system collects 3D data points with an (x,y) spatial resolution of 1 mm it is not possible for the system to identify individual particles close to this spatial resolution. As it is necessary to be able to identify individual particles, and small parts of overlapped particles, it is reasonable to expect that the system cannot detect individual particles below 5 mm. Therefore a data set comprising particles in the range 0–2 mm was obtained to ensure that there should not be any particles large enough for the system to identify. Measurements were performed on stockpiled products in the size range 0–2, 10–25, 20–40, 40–70, 60–90 mm, and then on three data sets from a primary crusher with material in the range 0–250 mm.

2.2 *Particle delineation*

The first step is to perform the detailed particle delineation (image segmentation) to identify the individual rock particles. These techniques have been applied to laboratory rock piles (Thurley & Ng 2005), in an industrial pellet measurement system

(Thurley & Andersson 2007) and rocks on conveyor (Thurley 2011), the latter of which forms the basis of the presented results. The technique is predominantly based on morphological image processing (Dougherty & Lotufo 2003, or any broad textbook on image processing), based largely on various edge detection techniques to facilitate seed formation for the watershed segmentation algorithm (Beucher & Meyer 1992, Dougherty & Lotufo 2003).

Figure 3 shows images of the 3D surface profile of rocks on the conveyor for the 20–40 mm data set, and 3 sets from the 0–250 mm crusher feed.

Figure 4 shows the automated particle delineation results. Particles have been removed from the delineation on each end of the conveyor in accordance with the *sampling and delineation strategy* suggested by Pitard (1993), removing all particles whose centre-of-mass is not placed within a defined transverse delineation zone.

Figure 3. 3D surface profile of rocks on conveyor. 20–40 mm product (top left), and three data sets from the output of a primary crusher in the range 0–250 mm.

Figure 4. Fully automated particle delineation. 20–40 mm product (top left), and three data sets from the output of a primary crusher in the range 0–250 mm.

The particle delineation is processed further to identify areas-of-fines (Thurley 2009), and non-overlapped particles (Thurley & Ng 2008). These methods work by following the perimeter of every delineated particle and examining the depth variation in the 3D surface data. Regions that are non-overlapped will be consistently above their neighbour regions. Regions that are rocks will typically have large depth variations around the perimeter, whereas regions that are fines will typically have small depth variations. This is because the areas-of-fines are typically over-segmented (contain many seemingly random regions) and the areas-of-fines have slowly varying surface topology (changes in depth).

Figure 5 illustrates this capacity to overcome *sub-resolution particle error* and automatically detect areas-of-fines, with the 0–2 mm data set shown on the left, with the regions from the delineation that were classified as areas-of-fines shown on the right in shaded/coloured blobs. This ensures that areas-of-fines are not mis-sized as large particles. There are regions on the left and right edges of the right data set that were not classified as fines. The data in these areas appears to have some vertically oriented ridges in the data that look significantly like rock structures and so are not detected as fines. An additional data set where the belt was more heavily loaded with the 0–2 mm product did not have these ridges and when analysed 98% of cumulative sieve-size distribution was classified as below 5 mm. Unfortunately this data was lost on a stolen computer and at the time of writing additional 0–2 mm data sets have not been collected.

Figure 6 shows all particles automatically identified as non-overlapped and illustrates the capacity to overcome *overlapped particle error* preventing mis-sizing of overlapped particles as small size classes.

Being able to largely eliminate these two sources of error removes error that biases the results towards smaller sizes (*overlapped particle error*) and error that biases towards large sizes (*sub-resolution particle error*).

Figure 5. 3D surface profile data and automated delineation for 0–2 mm particles, showing large regions detected as areas-of-fines.

Figure 6. Automatically identified non-overlapped particles. Left image corresponds to Figure 3 bottom left. Right image corresponds to Figure 3 bottom right.

A system with opposing sources of error will tend to be unstable and lack the capacity to trend in the right direction. As we can eliminate these sources of error, the presented research has been consistently demonstrated to trend in the right direction.

2.3 *Particle classification & sizing*

Particle sizing represents a significant improvement in this research to previous work. A new volumetric based sizing has been applied that uses the 3D profile data of each particle, and as a result no "calibration" or statistical matching to expected sieve results was performed in this research.

Particle sizing is performed in the following way;

1. Each region in the particle delineation is classified as fines, overlapped, or non-overlapped based on the methods published earlier by Thurley & Ng (2008) and Thurley (2009).
2. Each region identified as fines is converted into a volume based on the 3D surface profile and the depth of the material on the belt at each point in the 3D surface profile (from a measurement of the weighted-down empty conveyor). The volume beneath the region is assumed to be fines. The fines volume is allocated to a particle size of less than 5 mm as this is the predefined minimum size in these cases.
3. Each region identified as overlapped is currently ignored. These regions are analogous to ice-bergs and are a more difficult prospect for sizing. Identifying and ignoring these particles prevents mis-sizing them are smaller particles.

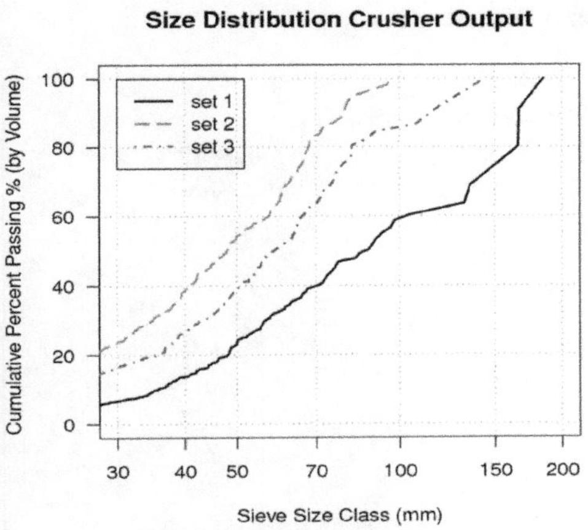

Figure 7. Calculated cumulative sieve-size distribution for the three crusher data sets in Figure 3. Set 1 (solid line) is the top right image in Figure 3, set 2 (dashed line) is the bottom right image, and set 3 (dot-dash line) is the bottom left image.

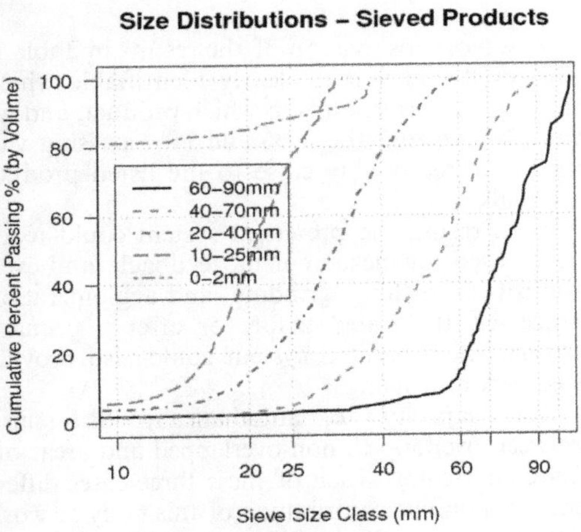

Figure 8. Calculated cumulative sieve-size distribution for the stockpiled products. From the left most distribution curve the product size is 0–2, 10–25, 20–40, 40–70, and 60–90 mm on the far right.

4. Each region identified as non-overlapped is processed as follows;
   a. The best-fit-rectangle in the horizontal plane is calculated for that region. From the length and width of this rectangle a major axis and a minor axis for the particle is defined, denoted *major*, *minor*.
   b. The 3D surface profile of the particle is analysed to calculate the maximum height, denoted Z_{max} and minimum height of the particle, denoted Z_{min}
   c. The height of the particle H is estimated to be double the known height of the particle, unless this height would make the particle extend below the conveyor belt. The height is approximated by Equation 1 as the minimum of the following two expressions where D_{zmin} is the depth of the material on the conveyor at the location of Z_{min}

$$H = \min[2 \cdot (Z_{max} - Z_{min}), D_{zmin} + Z_{max} - Z_{min}] \quad (1)$$

   d. The volume of the particle is approximated using an ellipsoid with axes *H, major, minor*. Therefore the volume V = (4/3)·π·*H*·*major*.*minor*
   e. The sieve-size of the particle is approximated as the minimum value of *H* and *minor*. Note that it would be more appropriate to estimate the sieve-size as the side length of the smallest square that can contain an ellipse with *H, minor* as its two principle axes and we will use this in the future.
   f. All particles with a sieve-size less than 5 mm are added to the fines volume.
5. The sieve-size-distribution is calculated by ordering all of the particles by their sieve-size (starting with the fines volume) and calculating the cumulative sum of all of their volumes. Figures 7 and 8 graph these sieve-size-distributions, illustrating the lack of predefined sieve-size-classes in the graphs. This allows any arbitrary sieve-size-classes to be determined after the analysis. Table 1 shows the 20%, 50% and 80% passing values for the curves in both figures.

Performing the analysis on over 6 hours of ship loading data in which the 40–70 and 60–90 mm

Table 1. Calculated 20, 50, and 80 percent passing values (mm) for each product.

Sieve-size (mm)	0–2	10–25	20–40	40–70	60–90	0–250		
80% passing	5	26.2	40.7	67.8	96.1	165	67.8	81.5
50% passing	5	20.5	31.4	55.0	76.0	84.1	47.9	58.1
20% passing	5	16.2	22.5	37.4	62.5	47.9	27.0	35.1

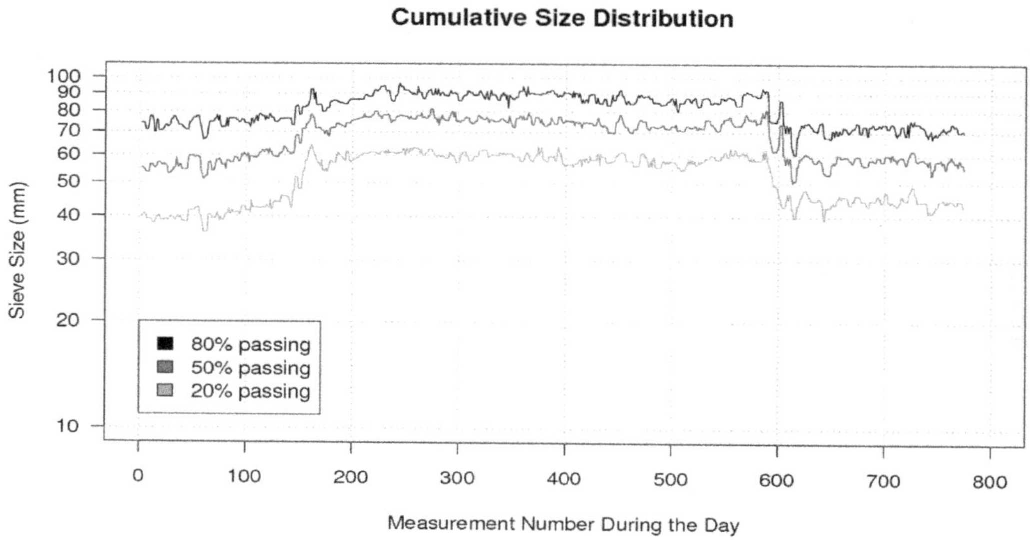

Figure 9. Over 6 hours of measurement data (767 measurements) during loading of the 40–70 mm and 60–90 mm products.

products were being loaded produces the graph shown in Figure 9. Only the 20, 50, and 80% passing values are plotted and one can clearly see when the loading changes and identify the product being loaded at any given measurement based on the 20% and 80% passing values.

3 DISCUSSION

3.1 *System results*

The presented research demonstrates a non-contact, automated particle delineation and analysis technique that uses 3D surface profile data on a broad range of material sizes and applies a new sizing strategy to directly calculate the sieve-size-distribution by volume (approximating weight). This allows the capability to install the system without any need to calibrate against manual sieve samples.

The results show a strong relationship to the listed product size range for the pre-sieved products and it is clear from observation of the 3D data in Figure 3 and the cumulative size distributions in Figures 7 and 8 that the results trend in the right direction. That is, when large material is on the belt, the size distribution result is larger.

Comparison against manual sieving results is not yet available but will be performed in future work to fine-tune the size estimation methods and investigate the system accuracy.

Based on observation of the results in Table 1, Figures 8 and 9 it is clearly identifiable which sieve curve corresponds to which product, and we can observe that the 20% and 80% passing values come reasonably close to the listed product intervals.

As a result, the presented system could readily be used for measurement, feedback, and even control of crushing, grinding, and agglomeration processes. If located before or after a primary crusher the presented system could also provide feedback to blasting.

The capacity to automatically distinguish between overlapped, non-overlapped and areas-of-fines, and to treat each of these three cases differently is a unique contribution of this body or work and it allows a number of sources of error to be addressed.

3.2 *Sources of error*

The following summary identifies how these sources of error have been mitigated, addressed, or ignored;

Particle delineation error is not the subject of this paper, but a detailed investigation is performed by Thurley & Ng (2005). One can note however, that using 3D surface profile data eliminates particle delineation problems due to shadows, and colour variation in the material which can affect photographic 2D systems.

Sub-resolution particle error is mitigated by this implementation as areas-of-fines can be automatically detected and sized as fines instead of being mis-sized as large particles.

Segregation and grouping error remains a constant source of error but appears not to be particularly significant for the stockpiled products. It may have a significant but as yet unmeasured impact on the case of the crusher material but as noted above, these results clearly trend in the right direction allowing the possibility of process control.

Overlapped particle error is entirely mitigated by the presented system as it prevents mis-sizing of overlapped particles as smaller particles.

Capturing error is ignored but could be further investigated in future work if increased accuracy is necessary.

Profile error remains a relevant source of error, as it is not clear how accurate our 3D volumetric estimation is for each individual particle. However, when the sample is considered as a whole, the size-distribution results suggest the *profile error* is not a significant problem. Further investigation could be performed in the future to evaluate and improve accuracy.

Sample delimitation and extraction is not a significant source of error as a correct delineation and extraction based on centre-of-mass is performed.

Weight estimation is mitigated in this implementation as volumetric estimation of each particle is performed. The assumption of constant density within a sample seems reasonable but will lead to a source of error. Weight estimation in this case reverts back to *profile error* and the correct estimation of the particles actual 3D volume.

Surface analysis is by definition a source of error. There is no avoiding this error, but there is the possibility to mitigate it further. Future work will include collecting the particle size information and bulk weight information from belt scales. With this information it may be possible to estimate the proportion of hidden fines in the material given the observed particle size distribution and any discrepancy between expected weight and measured weight.

4 CONCLUSION

This research demonstrates algorithms and a measurement system for automated, non-contact, on-line, particle size measurement that can be used without the need to calibrate or statistically correct the measurement results against a priori sieving results. Moreover the research demonstrates how relevant sources of error are addressed and mitigated in order to ensure a system that can be reliable and trend in the right direction. Without mitigating these errors, particularly *overlapped particle error,* and *sub-resolution particle error*, it seems unrealistic to expect reliable results. Key outcomes include;

1. The core capabilities to;
   a. automatically delineate particles,
   b. classify delineated regions as overlapped particles, non-overlapped particles, or areas-of-fines, and
   c. size particles based on volume in order to estimate the sieve-size-distribution
2. Sieve-size-distribution results that are close to expected product sizes,
3. Results that trend in the right direction,
4. Results that are clearly appropriate for feedback and control of industrial processes, and
5. A system that can work without any calibration against manual sieving making it easy to setup and maintain.

The presented algorithms are readily applicable to other mining processes, such as the feed hopper to a primary crusher, or in excavator buckets, shovels, drawpoints, and muckpiles, using an alternate appropriate method of collecting 3D surface profile data in that environment.

Some further development and testing is planned including;

1. Large scale testing collecting weeks of production data and comparison to daily manual sieving results.
2. Estimation of hidden fines may be possible based on the presented technique coupled with on-line weight information from belt scales.
3. Improvements in computational speed so that measurements can be performed in a few seconds or faster in order to allow fast automatic control strategies.
4. Improvements and tuning of the sieve-size calculation.

ACKNOWLEDGEMENTS

The author would like to thank all those who have supported this research over the last 15 years. It is the beginning of the end of the image analysis research effort for this method, and sincere thanks go to all those who have been involved, specifically; Luleå University, ProcessIT Innovations, Nordkalk, MBV-Systems AB, MinBaS, LKAB, SSAB, New Boliden, VINNOVA, EU INTERREG IVA Nord, Softcenter AB, CSIRO Exploration & Mining, Monash University, Sven Molin, Olov Marklund, Pär-Erik Martinsson, Anders OE Johansson, Håkan Pihl, Marianne Thomaeus, John-Erik Larsson, Jan Nyström, Peter Wedin, Tobias Andersson, Johan Carlson, Kim Ng,

John Minack, Cathy Wilkinson, Bruce Hobbs, Alison Ord, Lawrence Cheung, David LeBlanc, Paul Maconochie, Anthony Maeder, and family, and to those that have been missed, sincere apologies.

REFERENCES

Andersson, T. & Thurley, M.J. 2008. Visibility classification of rocks piles. *Proc. 2008 Conference of the Australian Pattern Recognition Society on Digital Image Computing Techniques and Applications (DICTA 2008)*. Australian Pattern Recognition Society, pp. 207–213.

Beucher, S. & Meyer, F. 1992. *Mathematical Morphology in Image Processing*. New York: Marcel Dekker. Chapter 12, pp. 433–481.

Carlsson, O. & Nyberg, L. 1983. A method for estimation of fragmentation size distribution with automatic image processing. *Proc. First International Symposium on Rock Fragmentation by Blasting—FRAGBLAST*, Luleå, Sweden, August 1983, pp. 333–345.

Cunningham, C.V.B. 1996. Keynote address: Optical fragmentation assessment—a technical challenge. In J.A. Franklin & T. Katsabanis (eds.), *Measurement of Blast Fragmentation—Proc. FRAGBLAST 5 Workshop*. Balkema, pp. 13–19.

Dougherty, E.R. & Lotufo, R.A. 2003. *Hands-On Morphological Image Processing*. SPIE—The International Society for Optical Engineering, vol. TT59.

Fernlund, J.M.R. 1998. The effect of particle form on sieve analysis: a test by image analysis, *Engineering Geology* 50: 111–124.

Franklin, J.A. & Katsabanis, T. (eds). 1996. *Measurement of Blast Fragmentation—Proc. FRAGBLAST 5 Workshop*. Rotterdam: Balkema.

Frydendal, I. & Jones, R. 1998. Segmentation of sugar beets using image and graph processing. *ICPR 98 Proc. 14th International Conference on Pattern Recognition, vol. II, Brisbane, Australia, August 1998*, pp. 16–20.

Kim, H., Haas, C., Rauch, A. & Browne, C. 2003. 3D image segmentation of aggregates from laser profiling. *Computer Aided Civil and Infrastructure Engineering*, pp. 254–263.

Lee J., Smith M., Smith L. & Midha, P. 2005. A mathematical morphology approach to image based 3D particle shape analysis. *Machine Vision and Applications* 16(5): 282–288.

Noy, M.J. 2006. The latest in on-line fragmentation measurement—stereo imaging over a conveyor. *Proc. Eighth International Symposium on Rock Fragmentation by Blasting—FRAGBLAST 8*, May 2006, pp. 61–66.

Ord, A. 1988. Real-time image analysis of size and shape distributions of rock fragments. *Explosives in Mining Workshop, Melbourne, Victoria, November 1988*. The AusIMM, pp. 115–119.

Pitard, E. 1993. Pierre Gy's sampling theory and sampling practice: heterogeneity, sampling correctness and statistical process control. CRC Press.

Potts, G. & Ouchterlony, F. 2005. The capacity of image analysis to measure fragmentation, an evaluation using split desktop. Swebrec—Swedish Rock Breaking Institute, Technical Report, ISSN 1653-5006.

Rosato, A., Strandburg, K., Prinz F. & Swendsen, R. 1987. Why the brazil nuts are on top: Size segregation of particulate matter by shaking. *Physical Review Letters* 58(10): 1038–1040.

Thurley, M.J. 2002. Three dimensional data analysis for the separation and sizing of rock piles in mining. Ph.D. dissertation, Monash University, December 2002.

Thurley, M.J. 2009. Fragmentation size measurement using 3D surface imaging (in lhd buckets). *Proc. 9th International Symposium on Rock Fragmentation by Blasting—FRAGBLAST 9, Granada, Spain, September 2009*, pp. 133–140.

Thurley, M.J. 2011. Automated online measurement of limestone particle size distributions using 3d range data. *Journal of Process Control* 21(2): 254–262.

Thurley, M.J. & Andersson, T. 2007. An industrial 3d vision system for size measurement of iron ore green pellets using morphological image segmentation. *Minerals Engineering* 21(5): 405–415.

Thurley, M.J. & Ng, K. 2005. Identifying, visualizing, and comparing regions in irregularly spaced 3D surface data. *Computer Vision and Image Understanding* 98(2): 239–270.

Thurley, M.J. & Ng, K. 2008. Identification and sizing of the entirely visible rocks from segmented 3d surface data of laboratory rock piles. *Computer Vision and Image Understanding* 111(2): 170–178.

Wang, W.X. & Stephansson, O. 1996. Comparison between sieving and image analysis of aggregates. In *Measurement of Blast Fragmentation—Proc. FRAGBLAST 5 Workshop*. Rotterdam: Balkema, pp. 141–148.

Measurement and Analysis of Blast Fragmentation – Sanchidrián & Singh (Eds)
© 2013 Taylor & Francis Group, London, ISBN 978-0-415-62140-3

Surfometric imaging for dimensional analysis of individual aggregates and bulk samples

G. Dislaire, P. Illing, C. Laurent & E. Pirard
Liège University, Belgium

P. Di Carlo & C. Moitroux
CECOTEPE, HEPL, Liège, Belgium

ABSTRACT: Aggregate size analysis in the mineral industry is still dominated by manual sampling and subsequent sieving. Online measurements are of interest to allow for optimization of the process with a real time feedback to the plant. Such measurements can be performed using conventional video cameras or by way of surfometry based on laser triangulation. This study investigates the measure of the size and the volume of stones by surfometry. This investigation is done in the case of individualized fragments and in the case of bulk samples of aggregates. The advantage of surfometry is the accuracy and speed of the measurement process, but the principal drawback is the lack of data relative to the hidden (lower hemisphere) part of the object. In order to increase the accuracy of volume estimation, an assumption about symmetry of the object has to be made. Volumes of individualized stones estimated from the surfometry measure are correlated to the actual volumes. This allows validating the hypothesis while it also highlights some limitations. Stones placed in a monolayer approximate the real condition on the conveyor belt. Estimated volumes are again correlated with the actual one. Both correlations are compared and differences are discussed.

1 INTRODUCTION

Size analysis of aggregates and crushed ore material is still dominated by manual sampling and subsequent sieving. The development and improvement of online measurements is of interest to allow the optimization of the process which provides a real time feedback to the plant. Such measurements can be performed using conventional video cameras or by way of surfometry based on laser triangulation (Fig. 1).

The present study investigates the measure of size and volume of individual stones by surfometry. Surfometric imaging provides a grey scale image wherein the grey level is directly related to the height of the surface at pixel position.

The advantage of using surfometry as well as other Computer Vision (CV) techniques is the accuracy and speed of the measurement process. Surfometric imaging is far superior to conventional camera acquisition for the following reasons, most of them already underlined by Thurley (2005):

- First, the topographic data allows for easier segmentation of individual stones based on the relative altitudes instead of having to outline stone boundaries based on a poorly reliable shadowing effect. Non-uniformity of lighting as well as rock textures often induce a bias in the detection, that surfometry easily overcomes. As a result, there is no need to fine tune the algorithm depending on the product texture (nature of the rock or ore).

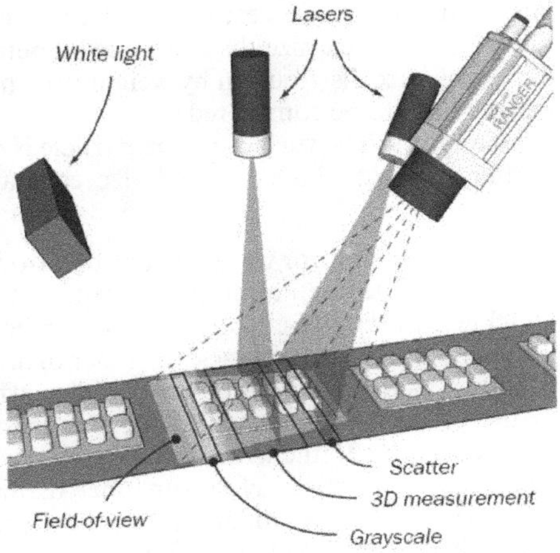

Figure 1. Principe of a laser triangulation acquisition system illustrating the geometry between the camera viewing axis and the incident laser plane (Doc. SICK).

- Second, using simple trigonometry, the stone mean altitude value gives direct access to a size calibration factor which has to be applied to it. Indeed, without size correction, stones on top of the conveyed heap are seen bigger than stones lying further from the sensor.
- Third, taking into account overlapped particles introduces a bias towards the smaller size classes. This can be addressed by surfometry using a visibility ratio as suggested by Thurley & Ng (2008). This parameter is based on the relative depth of pixels in the neighborhood of the particle boundary. Thurley concludes that "*the best-fit-rectangle of the entirely visible rocks provides a measure of size that correlated with the actual sieve size*".

Other sources of errors listed by Thurley or by Andersson (2010) are:

- Segregation and grouping errors: Due to vibration, large particles tend to move toward the top of the heap. Those errors can't be compensated by surfometric information and will need a weighting model to transform the surface size distribution into a pile size distribution.
- Capturing errors: Bigger particles are more likely to hit the frame of the imaging window. This error is reduced by removing the overlapped particles. Indeed bigger particle are also more likely overlapped.
- Weight transformation errors: In order to transform from particle size distributions by number to particle size distribution by weight, weighting models have to be considered.
- Profile errors: The surface of the particle is only partially visible which induced size estimation errors.

Another kind of error we would like to introduce here is the particle orientation error, or the fact that the thickness of the stone is often not perpendicular to the belt plane. In those cases, the inner diameter also known as the maximum inscribed disc, which is in the plane perpendicular to the thickness, is not perpendicular to the laser scan line, producing a smaller 2D projection. However inner diameter also known as the inner diameter presents properties described by Pirard et al. (2004).

2 METHODS

2.1 *Definitions*

In order to explain the orientation error, we must first define what are the stone dimensions and what are the sizes measured by computer vision.

A stone can be described by its length—the longest dimension, its thickness—the smallest dimension, its width—the smallest dimension measured with a caliper perpendicular to the length and thickness, and its sieve size that is the sieve opening diameter—the smallest round mesh diameter allowing the stone to pass through it.

Measuring those dimensions by computer vision requires the camera (2D) or the laser line (3D) to be perpendicular to them. It is good practice in 2D particle Static Image Analysis (SIA) to acquire the width-length plane. Indeed, this one includes the width, the length and the maximal inscribed disc, which is strongly correlated to the sieve size.

This last plane has also the property to be almost parallel to the belt plane when the particle is in a position of rest on this one. For this reason we will name it the rest plane in the following paragraphs.

Figure 2 illustrates the width-length box also known as the best fit rectangle and the inner diameter computed by image analysis. Two definitions of the width-length rectangle exist. We used the one parallel to the inertia ellipsoid axes.

When measuring stones in heap, the width-length plane is not necessarily parallel to the belt plane as illustrated in Figure 3.

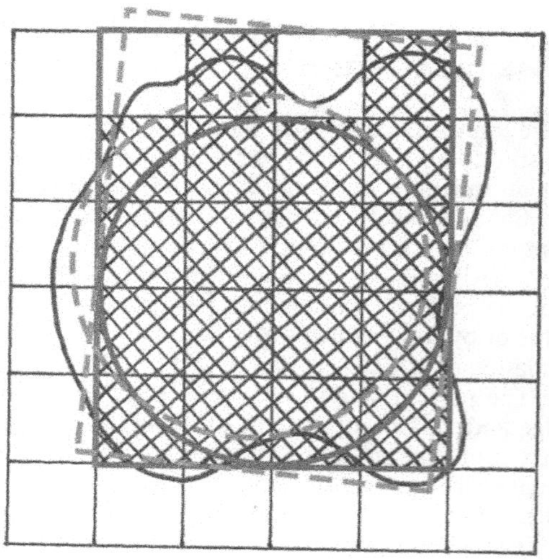

Figure 2. 2D particle projection in the rest plane including the length, width and maximal inscribed disc. The blue rectangle (continuous line) represents the width and length defined as the Feret Min diameter and the Feret diameter perpendicular to the Feret Min diameter. The red rectangle (dot line) represents the width and length defined as the Feret diameters parallels to the major and minor axes of the particle inertia ellipsoid. Those two rectangles provide very similar measures. In the case of 2D SIA with particle in a rest state, the rest plane is parallel to the belt plane and thus perpendicular to the camera.

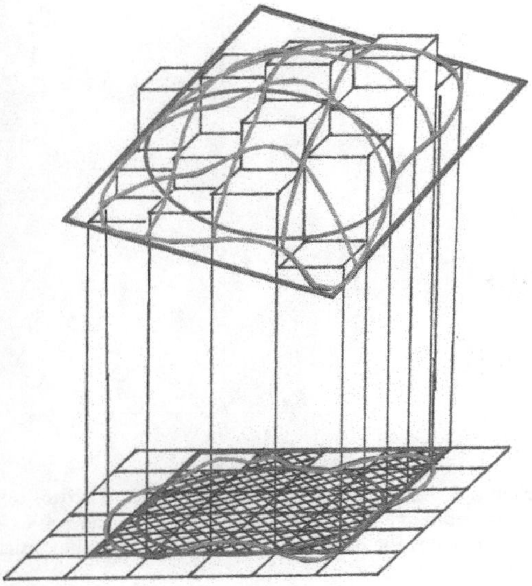

Figure 3. 3D surface pixels and corresponding 2D projected area on belt. When the rest plane is not parallel to the belt plane, the width-length rectangle projection on the belt plane defines Computer Vision (CV) width and CV length smaller than the stone width and length. The maximal inscribed disc within the projected area is also an underestimation measure of the sieve size.

In this case the width, length and maximal inscribed disc dimensions found within the 2D stone projected area on the belt plane are biased measure of those dimensions within the rest plane. This produces underestimations that can be corrected only with the determination of the rest plane.

We will estimate the rest plane as the regression plane passing by the stone boundary pixels. In this study, this plane has been used to estimate the volume as a ratio of the one included between it and the stone 3D surface. Nevertheless width and maximal inscribed disc have not been corrected using it. This would be an interesting improvement for forthcoming investigations. In this one, width has been corrected with the regression coefficient of the CV width vs. measured width plot illustrated further.

The idea is to reduce these errors as much as possible by making a clever use of the 3D information. Whereas the segregation error can't be compensated by making use of surfometry, the weight transformation error, the profile errors and the particle orientation errors can to some extent be reduced. These last errors and the problem of the volume estimation are closely related and are the main subject of this paper.

2.2 *Segmentation, classification, segregation correction*

As explained by Thurley & Ng (2008), the rock pile size distribution problem can be separated into the subproblems of segmentation, classification into size distribution and correlation between the surfometric size distribution and the pile size distribution.

The first subproblem is the only one to be mostly solved: A good segmentation algorithm can be visually validated. The one we implemented, is based on a technique similar to the one used by Thurley (2011, Fig. 4). An original edge detection technique is first applied, followed by a distance function to improve the location of appropriate markers for a watershed segmentation.

The third subproblem deals with the segregation errors. This is a statistical problem that Andersson et al. (2012) treats with training sets and a classifier. When the surface size distribution is correlated to the pile size distribution obtained from sieving, a model can be found to transform from one distribution to the other one.

Here, we would like to focus again on the second problem in a slightly different way than the one proposed by Thurley. Instead of using a best-fit-rectangle to estimate the stone size, we suggest the use of the inner diameter in the rest plane, which does physically correspond to the sieve size. As explained, it has been confirmed by Pirard et al. (2004) that this diameter is a direct measure of the sieve size. Its exact estimation needs the particle to be at the rest or with its thickness perpendicular to the belt plane.

Another difference is that we will make an estimation of the true particle volume using the 3D surfometric data and not just a 2D projection. This point is more critical. The estimation of the volume

Figure 4. Stones acquired in heaps are segmented by watershed starting from seeds centered on each stone. Those seeds are the maximums of a distance function built from filtered edges.

is very sensitive as it is difficult to estimate the hidden portion of a stone from only its visible part.

These two differences in the measure of the size and the volume tend to postpone as much as possible the use of conversion models from Image Analysis (IA) to physical measures in order to properly investigate the weight transformation errors, the profile errors and the orientation errors directly from IA estimation of the sieve size and the particle weight. In this paper, sieve size and volume estimation will be performed for individualized fragments and for bulk samples of randomly oriented aggregates.

2.3 *Material*

Hundred aggregates have been individually measured using a caliper for their principal axes and pycnometer to estimate their individual volume precisely. Surfometric images of these same hundred aggregates randomly positioned are then acquired (Fig. 5).

2.4 *Size calibration*

In order to adequately measure the sieve size from maximal inscribed disc, a size calibration has to be applied on each stone according to its altitude on the belt.

Let us consider Figure 6. With *MAG* for magnification = DC/AB = (*WD* – *FL*)/*FL*, *WD* for working distance is OD, *FL* for focal length. The angle α is ADZ.

In our setup, magnification of 17, working distance of 440 mm, and an angle of 30° are used.

Trigonometry allows computing the altitude z (distance DZ) from the distance AB:

$$z = \mathrm{DZ} = \frac{MAG(z) \cdot \mathrm{AB}}{\sin\alpha} \tag{1}$$

$$MAG(z) = MAG \cdot \frac{WD - z \cdot \cos\alpha}{WD} \tag{2}$$

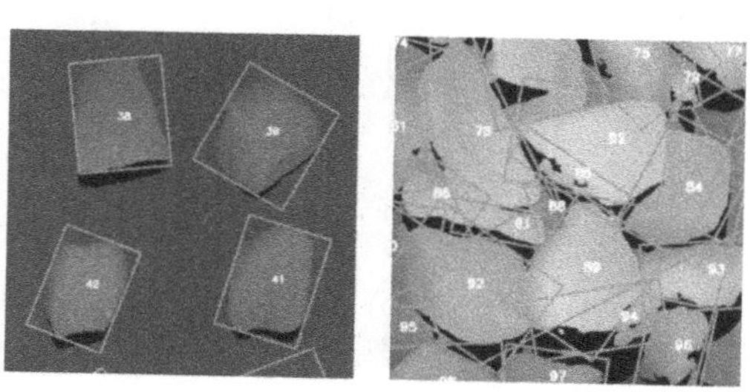

Figure 5. Stones are acquired as well individualized objects and in heaps. The computed Feret box and individual labels are overlaid on each stone. Note that only entirely visible stones will be kept and measured.

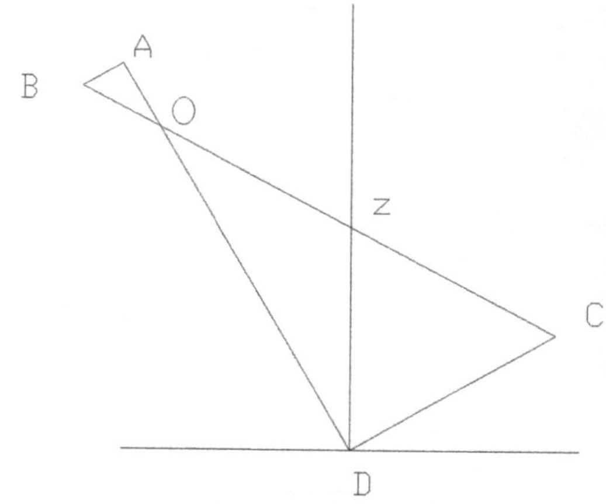

Figure 6. Schematic representation of the acquisition setup geometry within the plane vertical to the belt axis: ZD indicates the laser plane direction. AD is the optical axis of the camera.

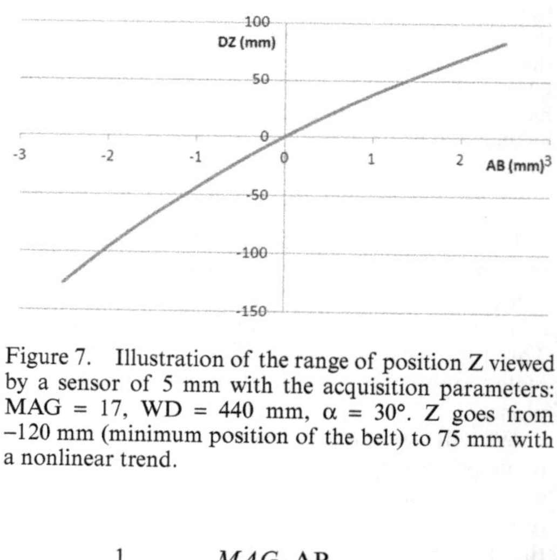

Figure 7. Illustration of the range of position Z viewed by a sensor of 5 mm with the acquisition parameters: MAG = 17, WD = 440 mm, α = 30°. Z goes from −120 mm (minimum position of the belt) to 75 mm with a nonlinear trend.

$$z = \frac{1}{1 + \dfrac{MAG \cdot \mathrm{AB}}{WD \cdot \tan\alpha}} \cdot \frac{MAG \cdot \mathrm{AB}}{\sin\alpha} \tag{3}$$

The variation of *DZ* compared to *AB* is non linear. See Figure 7. The most interesting thing is that the variation of Z depends of the altitude or, on the camera side, of the B position. Let's define the z resolution as $\Delta z = z_{j+1} - z_j$, where j is the camera pixel index in the AB direction.

A similar construction in the plane passing by BZ and perpendicular to the plane BZD (the plane of the Fig. 6) gives:

$$x = MAG(z) \cdot \mathrm{AB}' \tag{4}$$

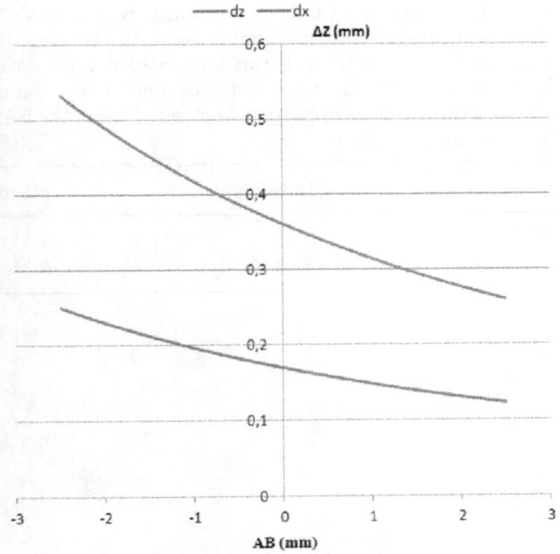

Figure 8. Object side pixel resolution compared to the object altitude. In our example, the grey values are in the range 0–255 for sensor B position in the range −1.28–1.28 mm. We see that a stone having a mean grey value of 32 (AB = −0.96) gives a Δz of 0.41 mm and a Δx of 0.19. The same stone with a mean grey value of 224 (AB = +0.96) gives a Δz of 0.31 mm and a Δx of 0.15. In those conditions and without calibration, a sphere of 10 mm will vary from 8.5 mm to 11.5 mm in size and thickness.

So, $\Delta x(z)$ varies like $\Delta z(z)$ with $\sin\alpha$ ratio. Figure 8 points out the variation of the pixel resolution depending on the stone altitude. Those variations are stronger when the camera angle decreases.

As $\Delta x(z)$ affects the measure of size (particle best-fit-rectangle size, width or inner diameter), a correct $\Delta x(z)$ calibration avoids binning a stone in an inadequate size class. In the same manner, as $\Delta z(z)$ affects the volume and therefore the weight estimation, only a correct $\Delta z(z)$ calibration avoids an inadequate weighting of the distribution.

Typical applications (angle of 30°, ratio field of view/altitude range = +/− 4) would present variations in size and volume up to 30% if a rigorous altitude calibration is not applied.

This calibration contributes to reducing the profile error, the orientation error and the weight transformation error.

2.5 *Size measured by surfometry*

In order to compare the size distribution from surfometric imaging to the one resulting from sieving, the inner diameter is the most adequate parameter. This parameter is easily computed from the projected area parallel to the belt, like other size parameters such as the length and width. Nevertheless, as in our set of stones, the manual measure included only length, width, thickness and volume, we studied the width parameter instead of the inner diameter.

Since the width measure suffers from the same limitations as the inner diameter ones, the correlations studied on this parameter should be similar to the one that would have been found on the latter.

The limitation is related to the fact that the stone has to present those sizes perpendicular to the laser incidence. In other words, the width and sieve diameter should be in a plane parallel to the belt plane.

When the stones are resting on top of a heap, their orientations depart from the belt resting position and both width and length are typically underestimated. The thickness is however overestimated.

2.6 *Volume measured by surfometry*

The volume measurement also suffers from the profile error, the orientation error and the weight transformation error. The volume has been estimated in this work using four different methods:

– First, an apparent volume has been estimated using the same principle as in 2D particle image analysis (Gregoire et al. 2007) when no information on the third dimension is available and only with A, the projected area.

$$Volume = \frac{4 \cdot A^{3/2}}{3 \cdot \pi} \quad (5)$$

– Second, the volume is estimated as the one of an ellipsoid having the same measured length, width and thickness as the major, median and minor axes.

$$Volume = \frac{\pi \cdot L \cdot W \cdot T}{6} \quad (6)$$

Here the thickness T is defined as the distance from the maximum z value to the regression plane.

– Third, the volume is estimated from the projected area multiplied by the thickness

$$Volume = \frac{2 \cdot A \cdot T}{3} \quad (7)$$

– Fourth, a volume is calculated by integration between the real 3D surface and a plane of regression fitted to the boundary points. This volume is doubled, making the assumption that the hidden surface is equal to the visible one.

3 RESULTS

Size (width) and volume of individualized stones estimated from the surfometry and calibrated. Each stone has been placed five times randomly on the belt and acquired. A good correlation between IA values and picnometry is illustrated in Figures 9, 10 and Table 1.

The correlations of the four volumes have been studied with R-squared of the linear regression values being respectively 0.23, 0.15, 0.26 and 0.33. As the integrated volume gives the best result only this one is reported.

In the case of stone placed on the belt, two integrated volume have been studied. One using neighbor points on the stone side, the second using neighbor point on the belt side. In the first case R-squared values are similar to the one found in heap.

In the last case, the regression plane corresponds to the belt. Figure 10 illustrates this last case where a very good correlation is found.

In the case of stone in heap, each stone has been dropped five time randomly on a stone layer. Figure 11 illustrates the CV volume and the physical volume correlation.

In this case, the R-squared value of the width linear regression is very bad. Nevertheless using the coefficient of the regression (=0.9) to correct the CV width value, we obtained a CV size distribution

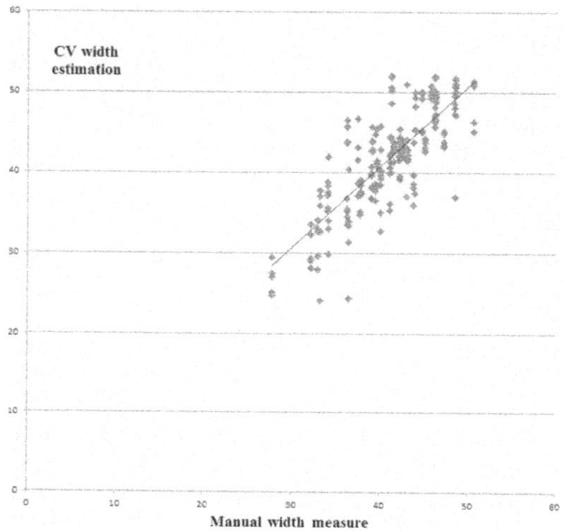

Figure 9. Image analysis width related to the manual width measured in the case of individual stones laid down on the belt.

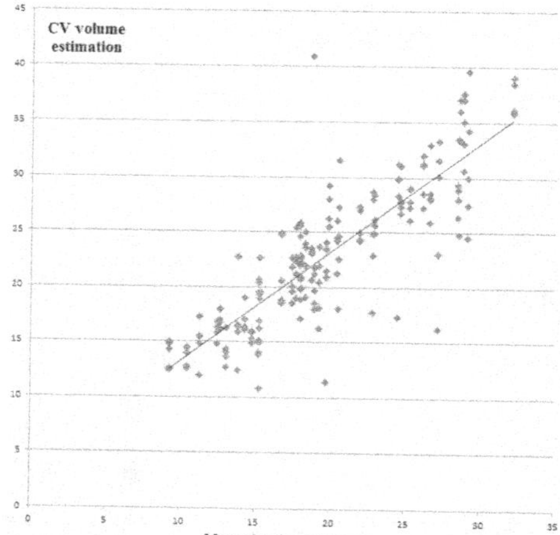

Figure 10. CV integrated volume related to the manual volume measured in the case of individual stones laid down on the belt.

Table 1. R-squared values of the linear regressions of the CV measures and the manual ones. In the case of individual stone on the belt two integrated volume have been studied; 0.33 corresponds to the similar regression plane as the one used in heap; 0.74 is found using the belt as the regression plane.

	Individual	Heap
Width	0.67	0.15
Volume	0.33 (0.74)	0.31

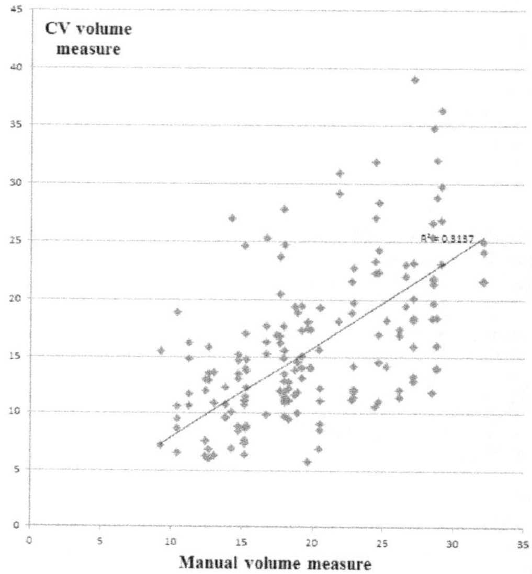

Figure 11. CV volume related to the manual volume measured in the case of stones in heap.

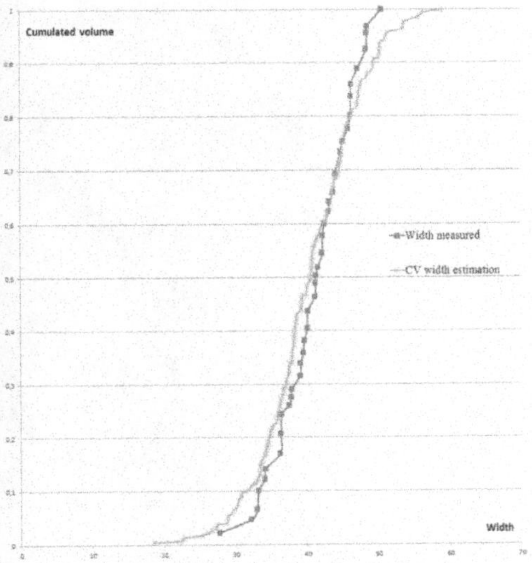

Figure 12. By weight CV size distribution compared to physical size distribution.

very similar the manual size distribution. This is illustrated in Figure 12.

4 DISCUSSION

In the case of individual measures, stones are in a resting state. This implies that the plane limiting from below the upper surface is not far from the plan including the median and major axis of inertia.

In case of measurement of stones in a monolayer this limiting plane is positioned in a more variable way. We examined the influence of this when measuring elongated stones. In this last case, we have seen that the volume estimated by surfometry is more robust than the size estimation. It could be interesting to correct the size with the knowledge of the regression plane. It could also be a solution to reject stones with regression plane too far from the horizontality.

5 CONCLUSIONS

The results offer many possibilities in terms of quality control and on-line optimization of the equipment.

The principal drawback of the surfometry is the lack of data relative to the invisible (lower hemisphere) part of the object. An assumption about symmetry of the object with regard to a bottom regression plane has been made. This method yields similar results to the ones obtained using the manual one.

This work focuses on selecting the computer vision methods that best match what is done physically rather than using and conversion table.

REFERENCES

Andersson, T. 2010. Estimating particle size distributions based on machine vision. Doctoral thesis. Luleå Tekniska Universitet.

Andersson, T., Thurley, M. & Carlson, J.E. 2012. A machine vision system for estimation of size distributions by weight of limestone particles during ship loading. *Minerals Engineering* 25(1): 38–46.

Gregoire, M., Dislaire, G. & Pirard, E. 2007. Accuracy of size distributions obtained from single particle static digital image analysis. *Proc. PARTEC2007—International Congress for Particle Technology (Nuremberg, Germany)*.

Pirard, E., Vergara, N. & Chapeau, V. 2004. Direct estimation of sieve size distributions from 2-D image analysis of sand particles. *Proc. PARTEC2004—International Congress for Particle Technology (Nuremberg, Germany)*.

Thurley, M. 2005. Identifying, visualizing and comparing regions in irregularly spaced 3D surface data. *Computer Vision and Image Understanding* 98: 239–270.

Thurley, M. 2011. Automated online measurement of limestone particle size distribution using 3D range data. *Process Control* 21(2): 254–262.

Thurley, M. & Ng, K. 2008. Identification and sizing of the entirely visible rocks from a 3D surface data segmentation of laboratory rock piles. *Computer Vision and Image Understanding* 111: 170–178.

Measurement and Analysis of Blast Fragmentation – Sanchidrián & Singh (Eds)
 ISBN 978-0-415-62140-3

Designing and optimising surface blasts using 3D images

M. Pötsch & A. Gaich
3GSM GmbH, Austria

R.A. McClure
RA McClure Inc., USA

ABSTRACT: A comprehensive knowledge of the blast site is for a controlled rock fragmentation essential. 3D images provide important information for the design and optimisation of blast site and the determination of structural geologic information. 3D images inherently provide combined geometric and visual information at a high level of detail and spatial accuracy. This paper addresses the generation of 3D images from a stereoscopic image pair, the registration to local and georeferenced co-ordinate systems, and merging of overlapping 3D images. A procedure for blast design and optimisation using 3D images is proposed including proactive design (before drilling) and integration of as-drilled data and updated 3D images. The procedure takes advantage of the minimum burden concept whose implementation is straightforward with a complete 3D image. 3D images are also used to determine parameters of the rock mass's fracture system allowing measurements at an arbitrary number and also in inaccessible areas. The data is used to characterise the rock mass and estimate the in situ block size distribution of a blast site.

1 INTRODUCTION

Comprehensive knowledge of the blast site geometry and rock mass properties supports to reduce some of the fundamental problems in surface blasting: (i) unexpected fragmentation, (ii) excessive vibrations, and (iii) fly-rock. Improving post-blast fragmentation requires an integrated control cycle including blast design, blast execution and fragmentation analysis (Raina et al. 2009).

Comprehensive knowledge of the blast site includes a detailed three-dimensional model of the bench face geometry preferably enhanced by visual information on the rock surface, and optimally data on the actual fracture network, all of these available before blasting.

A 3D image combines visual and geometric information of a rock surface hence being suitable to provide such comprehensive knowledge for bench face surveying. Although the technology and algorithms date back 150 years (Photogrammetry, e.g. Slama 1980) and 30 years (Computer Vision, Faugeras, 1993), respectively, practical systems were introduced mainly after year 2000 (Roberts & Poropat 2000, Gaich et al. 2004, Birch 2008).

In the meantime digital camera technology, mobile computing and graphics power, as well as algorithmic improvements (Bradski 2000) made 3D images a mature and well-proven method (Haneberg 2006, Moser et al. 2006, Tonon & Kottenstette 2007, Gaich et al. 2008).

This paper addresses briefly the use of 3D images for generating a geometric and visual 3D model of a blast site and shows how to use this data for designing and optimising surface blasts. It is also shown how the fracture system of a rock mass is measured from 3D images. These data are used to characterise the rock mass with respect to fracture sets, fracture orientations, fracture spacing, and in situ block size distribution.

2 GENERATION OF A 3D IMAGE

A three-dimensional image combines a set of photographs and the three-dimensional geometry of the imaged objects (surfaces). It thus combines visual with geometric information.

The creation process starts from at least a pair of (digital) images of the object of interest taken from different locations (stereoscopic image pair). A dense set of corresponding image points is derived automatically from these images utilizing techniques from multidimensional statistics (image matching). From the corresponding points the relative orientation of the cameras, i.e. the position and angular orientation of the cameras during imaging are automatically recovered (Hartley & Zisserman 2000). The corresponding points and the relative orientations allow to spatially intersecting rays (Fig. 1) leading to a set of points in 3D (point cloud) which are inter-connected to a

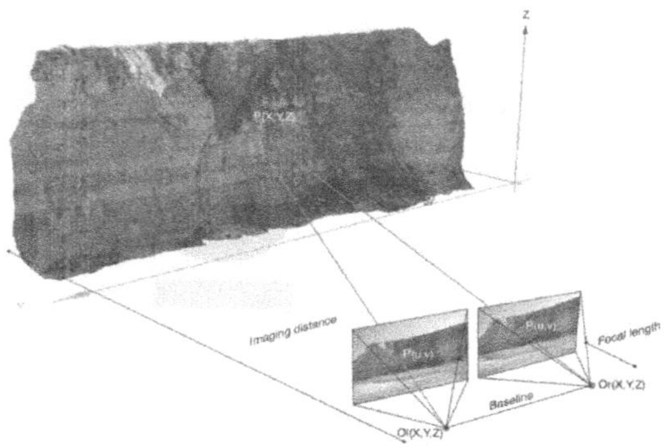

Figure 1. Principle of 3D image generation: from a pair of photos taken from different angles the geometry of the observed surface is automatically reconstructed. Modern algorithms do not need knowledge on imaging locations or the baseline.

Figure 2. Overlapping 3D images (top) and merged result (bottom). All underlying photos have been taken freehand without surveying any camera station and no markers are required in order to do the merge.

surface description (a mesh). The final 3D image is then formed by an overlay of the digital photograph on the irregular surface.

Though this principle works with photos from virtually any digital camera, knowledge on the camera and its lens is required to achieve geometrically accurate results. The required information—also called interior camera orientation—comprises: (i) the focal length, (ii) the direction of the optical axis, and (iii) parameters of the lens distortion. These parameters are determined by camera calibration.

Cameras can be calibrated before or after a measurement campaign. When a proper treatment of a camera is assured, cameras can be pre-calibrated thus reducing efforts during standard measurement activities.

Practical tasks often involve dealing with large areas, complex shapes or high resolution, each going beyond the information available from a single stereoscopic image pair. In these cases, the rock wall is acquired by several overlapping stereo pairs. They are automatically aligned to a larger 3D image based on common *structural and topographic* information in the overlapping area of the 3D images (Fig. 2). Note that no special reference points are required in order to link the images to each other.

Using *merged 3D images* it is possible to survey large blasts while sufficient image resolution is maintained to obtain a detailed geometric description of the bench face and details on the rock mass structure. Figure 3 shows a merged 3D image of a cast blast with six overlapping subimages in different colours. Merged 3D images allow also for the capture of open corners which is inherently not possible with a single stereo pair.

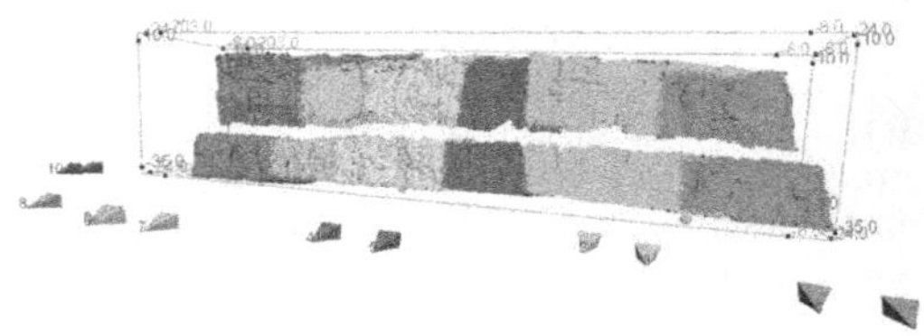

Figure 3. Larger area in an open pit captured by several overlapping 3D images and according camera locations. This way cast blasts of over 800 m length and beyond are surveyed.

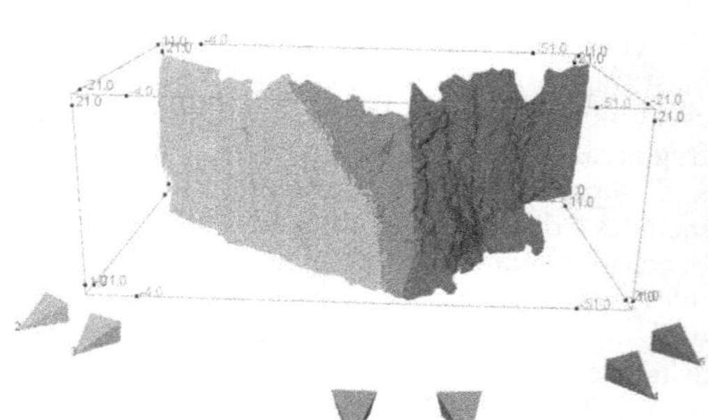

Figure 4. Open corner survey with a merged 3D images comprising three subimages.

Figure 4 shows an open corner surveyed by a merged 3D images which comprises three subimages.

3D images feature local and georeferenced co-ordinate systems. Using local co-ordinate systems blast design and fracture mapping can be performed without additional measurement devices just with a purpose-built set of reference markers to define scale, orientation, a borehole reference line, and an uneven crest. Alternatively, georeferenced co-ordinate systems are used to combine data from different sources such as total stations, or GPS. This integrates surveyed borehole collars and also surveyed crests.

3 3D IMAGES FOR BLAST DESIGN AND OPTIMISATION

The knowledge of the blast site geometry is of prime importance before doing the loading. The blast site geometry includes the bench face itself and its irregularities as well as the locations and trajectories of the boreholes.

A 3D image of the bench face enables the designer to plan and adapt the drill pattern according to the geometric irregularities of the bench face. Even if the boreholes have already been drilled, it is possible to account for the current burden situation when designing the loading.

Additionally, the actual rock mass conditions are represented in a natural way so it is straightforward to take the geologic situation at least qualitatively into account when doing the blast planning.

3.1 *General procedure*

The following procedure proposes a way how to use 3D images for designing and optimising surface blasts. The procedure provides several automatic tools but leaves the final design decision at the responsible blaster.

1. Generate a 3D image of the bench face before any boreholes are staked out or drilled
2. Specify the basic drill pattern on the computer by burden, spacing, inclination and azimuth of holes, sub-drilling, row shift
3. Inspect the drill pattern in conjunction with the bench face interactively in 3D and check burden information over the entire face (see section 3.2.1) and/or the minimum burden diagram (see section 3.2.2)
4. Relocate individual shot holes and/or change their inclination or azimuth according to bench face irregularities (see section 3.3)
5. Stake out the planned optimised drill pattern
6. Drill the holes
7. For comprehensive planning: survey location of collars and borehole deviations (see section 3.4); otherwise believe in the drilling and go directly to step 10
8. If the bench face has been concealed (e.g. by a muck pile), take stereo pairs of the excavated face areas and merge it with the original 3D image.
9. Update computer planned layout with surveyed borehole data and updated 3D image. This updates all profiles, sectional areas, and minimum burden information.
10. Use updated profiles/minimum burden diagrams for loading.

3.2 *Minimum burden*

The key information for the geometric optimisation according to steps 3 and 4 of abovementioned procedure is the real minimum burden. When using 3D images, information on minimum burden is available in two ways: (i) either visualised over the entire area of the blast or (ii) as minimum burden diagram analogously to conventional profiles.

3.2.1 *Minimum burden over the entire area*

Minimum burden over the entire area is achieved by determining the shortest distance between all surface points to the nearest borehole. By colourising the resulting distances according to the design burden and a corridor one gets an intuitive visualisation of the actual burden situation. Figure 5 shows an example for a three-step burden colourisation: green shows the areas at design burden plus/minus a definable corridor of acceptance while blue areas outline overburden and red highlights the areas of light burden. Since a 3D image usually provides surface measurements at some centimetres spacing on the bench, the resulting colour coded map reflects burden information at the same density over the entire surface.

3.2.2 *Profiles vs. minimum burden diagrams*

Profiles are generated by intersecting a plane with the free face of the bench and the boreholes. However, a profile does not necessarily show the shortest distance from the borehole to the surface. Especially for irregular faces containing cavities or at free ends and corners, a significant disparity between the profile and the shortest distances to the surface might occur.

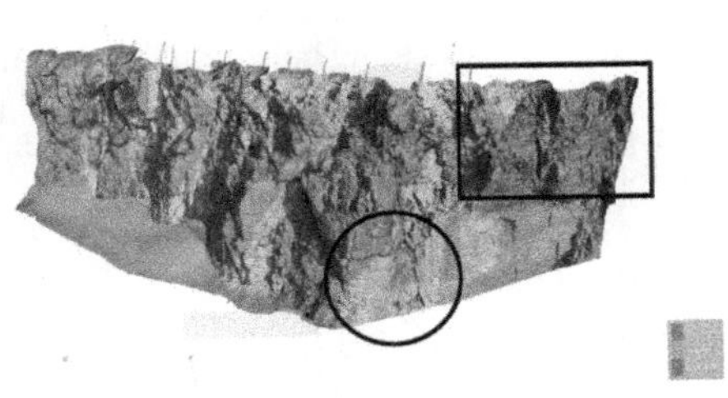

Figure 5. The colour-coded visualization of burden over the entire face facilitates the identification of light burden zones (circle) as well as regions with overburden (rectangle). In addition the underlying photos allow a visual estimation of weak zones.

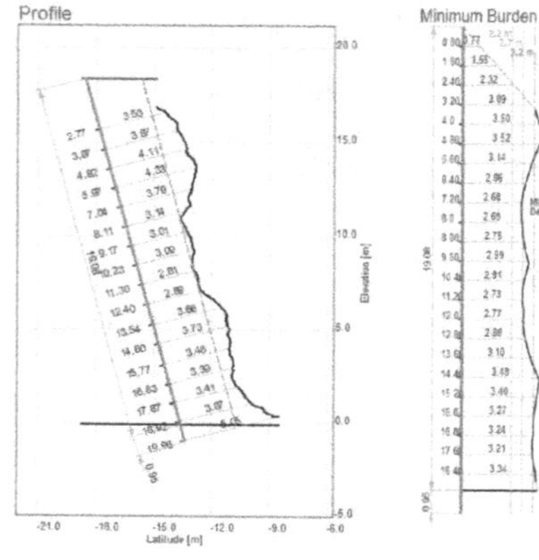

Figure 6. The profile and minimum burden diagram for the same borehole might show significant discrepancies. The profile naturally shows the larger burden values.

In contrast, the minimum burden diagram visualises the shortest distance from the hole to surface in any spatial direction at a certain depth. That means a minimum burden diagram depicts a 360° spherical search around the shot hole.

Figure 6 shows a profile and the minimum burden diagram for the same hole. The discrepancies at certain depths are in the range of several decimetres. Hence, light burden situations might stay unrecognised within a profile. Even when taking several profiles close beside each other, it is not guaranteed that the real physical minimum burden is revealed.

3.3 *Borehole relocation or adaptation of loading*

Once the real minimum burden information is available, there are two ways to adapt: (i) by relocating boreholes until the minimum burden fits the corridor of acceptance and/or (ii) by adapting the loading, e.g. using additional stemming at depths showing light burden. Any changes instantly update the information for the minimum burden diagram and the colour-coded burden visualisation as well. An optimal blast layout requires relocating holes until burden information is as expected which is a straightforward process when using 3D image technology. Figure 7 shows the possibilities for relocating boreholes.

3.4 *Borehole deviation data*

A complete knowledge of the blast site includes the true trajectory of boreholes. Significant deviations from planned to realised boreholes have been observed and analysed in the past (e.g. Reichholf 1997).

Several down-the-hole probes exist that allow for the determination the actual trajectory of the hole.

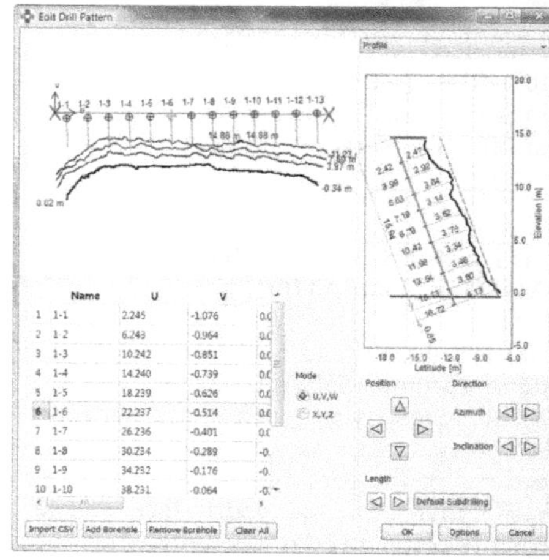

Figure 7. Every borehole can be individually adapted according the actual burden information given by profiles, minimum burden diagrams, and burden information over the entire face. Parameters include position, length, inclination, and azimuth.

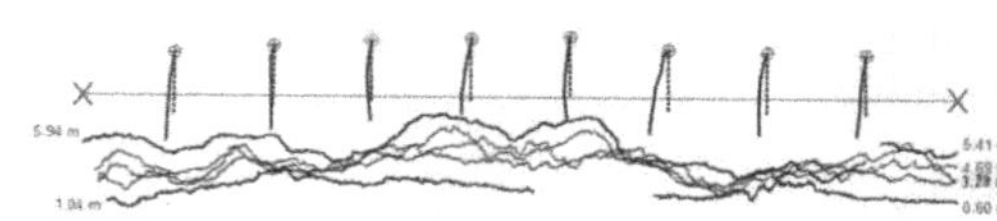

Figure 8. Plan view of a drill pattern with the contour lines of the bench. Planned holes are dashed lines while drilled hole trajectories are solid lines.

Figure 9. Comparison of the burden situation for the planned holes (top) and the drilled hole trajectories (bottom). Several light burden situations could be identified.

When combining their results with the geometric data of the bench face resulting from the 3D image, burden information is again updated leading then a complete geometric model of the blast site. Figure 8 shows a comparison between planned and drilled holes. Figure 9 shows the burden situation for the planned and drilled holes depicted in Figure 8. The holes as drilled led to several light burden situations in the bottom of the bench.

4 DETERMINATION OF ROCK MASS CONDITIONS

Aside of the blast parameters (drill pattern, loading, timing, etc.) properties of the rock mass play a significant role for the final fragmentation result (e.g. Scoble et al. 1996, Kilic et al. 2009). They include physical-mechanical properties of the rock types (Thuro 2002), and the properties of the fracture system. In particular, the orientation of fracture sets (Chakraborty et al. 1994) and in situ block size distribution has been found to have significant influence (e.g. Mathis 1995, Aler et al. 1996, Raina et al. 2009).

4.1 *Measurement of fractures*

Upon their combined visual and geometric information 3D images have been used for the acquisition of rock mass parameters, in particular geometric parameters of rock fractures (Gaich et al. 2006). With a purpose-made software component quantitative fracture parameters are gained including:

- Orientations from fracture surfaces (see Fig. 10) given by dip angle and dip direction
- Size, position, and shape of daylighting fracture surfaces
- Orientations of fractures from their corresponding traces at the surface (see Fig. 11) given by dip angle and dip direction
- Length and position of fracture traces
- Aperture (at high resolution)
- Termination
- Roughness (at high resolution) and waviness

Based on the set of single measurements it is straightforward to derive structural rock mass parameters such as:

- Fracture spacing and frequency (Fig. 12)
- Number of fracture sets (Fig. 13)
- Orientation statistics of fracture sets (Fig. 13)
- Fracture persistence and trace length distribution

Figure 10. Fracture orientation on a surface measured by dip angle and dip direction. The area encloses the daylighting part of a fracture plane and thus defines its position.

Figure 11. Fracture orientation measurement from its trace. The set of non-collinear points of the (irregular) fracture trace defines the orientation and position of the fracture plane.

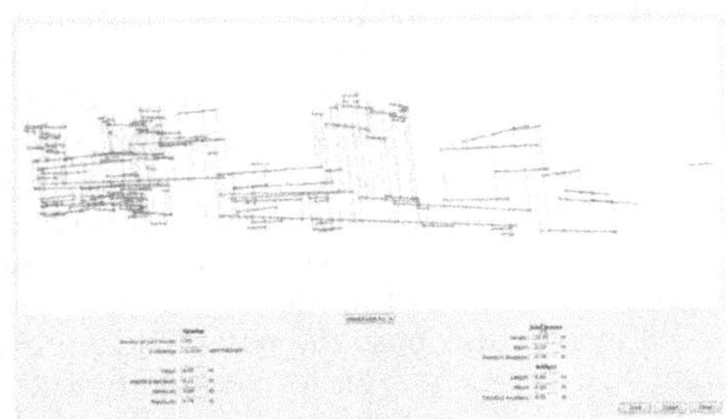

Figure 12. Sketch of projected traces forming the basis of the normal set spacing calculation.

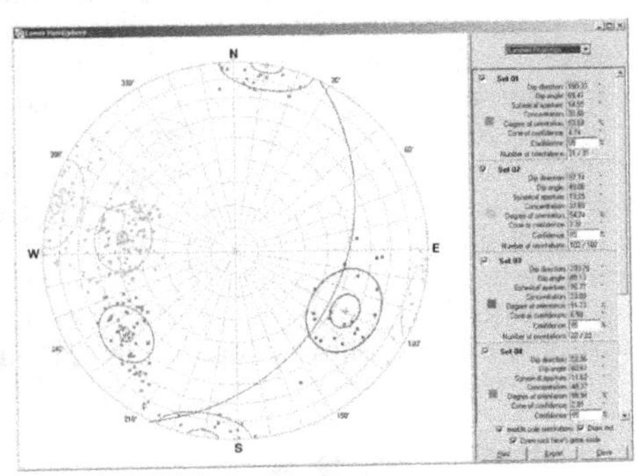

Figure 13. Hemispherical plot of fracture orientations grouped into fracture sets.

4.2 *In situ block size distribution*

The parameters assessed from a 3D image describe the fractures at the visible rock surface while the in situ block size distribution is a spatial parameter considering fractures inside the rock mass. Thus, it is only possible to obtain information about the block sizes by correlation with surface parameters or by fracture network modelling.

Palmström (2001) gives a correlation for the in situ block size based on the volumetric joint count *Jv* and a block shape factor β. He points out to calculate at least minimum, mean, and maximum block sizes. The volumetric joint count can be determined from the fracture spacing. Input parameters assessed from an area survey are preferred over a linear survey.

Several authors have proposed the use of discrete fracture network models (Fig. 14) to estimate the in situ block size distribution (e.g. Aler et al. 1996, Wang et al. 2003, Kim et al. 2007, Elmouttie & Poropat 2012). This approach requires the generation of discrete fracture, either persistent or with finite size, at deterministic and/or stochastically predicted locations, and the identification of finite blocks within the network. The identified blocks are used to calculate the block size distribution.

Both approaches, correlation and fracture network, benefit from high-quality input data obtained from 3D images as described in section 4.1. The data basis extends to a relevant volume of rock at a statistically significant number of measurements taken in a fraction of time compared to spot samples originating from traditional mapping methods. 3D images also pose the potential to provide automatic assessment tools for fast and objectively identifying planar fractures and traces (e.g. Falcao et al. 2000).

The drawback of the correlation method is that it is an empirical relationship, which is, although based on experience, subject of discussion. The drawback of the latter method is that it is computationally intensive and requires profound knowledge on fracture network modelling which could be impractical for a routine application in blasting. A current topic of investigation of the authors is the application of 3D images for determining the size distribution of surface patches of blocks. The investigations aim at correlating the patch size distribution with the in situ block size distribution for establishing a relationship for routine application.

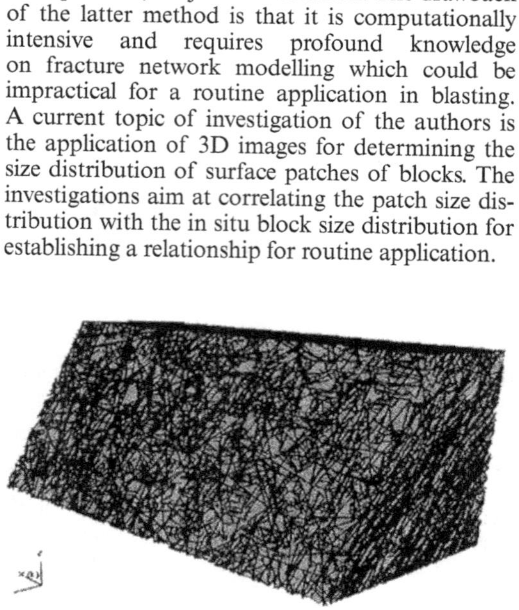

Figure 14. 3D view of a discrete fracture network of a rock slope (from Rogers et al. 2006).

5 CONCLUSION

The contribution presents a 3D imaging system used for blast design and optimisation, and measurements of rock fractures as well. It shows the generation of a 3D image from a stereoscopic image pair, registration to local and georeferenced co-ordinate systems, and merging of overlapping 3D images. Recent applications for large cast blasts and corner surveys are addressed. A procedure for blast design and optimisation is proposed which includes proactive design (before drilling), integration of as-drilled data and updated 3D images before loading. The procedure takes advantage of the minimum burden concept. 3D images are also used to determine parameters of the rock mass's fracture system allowing measurements at an arbitrary number and also in inaccessible areas. Descriptive parameters such as fracture orientations, spacing, persistence, or roughness coefficients can be gained directly from 3D images. The data is suitable as input for the determination of the in situ block size distributions, either with empirical relationships or discrete fracture network modelling.

The authors are convinced that a controlled rock fragmentation requires an integrated cycle including blast design, blast execution and blast result analysis. In this context an energy-controlled blast design has been promoted and its advantages shown by e.g. Moser (2003), Heck (2004), and Segarra et al. (2005). A complete 3D model of the blast site (bench face and boreholes) as well as the structural geologic situation (fractures) provides key information for the design process reducing several uncertainties with respect to bench face irregularities and rock mass conditions.

REFERENCES

Aler, J., du Mouza, J. & Arnould, M. 1996. Measurement of the Fragmentation Efficiency of Rock Mass Blasting and its Mining Applications. *Int. J. Rock Mech. Min. Sci. & Geomech. Abstr.* 33(2): 125–139.

Bradski, G. 2000. *The OpenCV Library*. Dr. Dobb's Journal of Software Tools.

Birch, J. 2008. Data acquisition with 3DM Analyst Mine Mapping Suite. In: Y. Potvin, J. Carter, A. Dyskin & R. Jeffrey (eds), *Proceedings of 1st Southern Hemisphere International Rock Mechanics Symposium*, Perth, 2008, 531–540.

Chakraborty, A.K., Jethwa, J.L. & Paithankar, A.G. 1994. Effects of joint orientation and rock mass quality on tunnel blasting. *Engineering Geology* 37: 247–262.

Elmouttie, M.K. & Poropat, G.V. 2012. A Method to estimate In Situ Block Size Distribution. *Rock Mechanics Rock Engineering* 45(3): 401–407.

Falcao, A.X., Udupa, J.K. & Miyazawa, F.K. 2000. An ultra-fast user-steered image segmentation paradigm: Live-wire-on-the-fly. *IEEE Trans. on Medical Imaging* 19(1): 55–62.
Faugeras, O. 1993. *Three-Dimensional Computer Vision.* MIT Press, Boston, MA.
Gaich, A., Schubert, W. & Pötsch, M. 2004. Reproducible rock mass description in 3D using the JointMetriX3D system. *Proc. of the ISRM Regional Symposium Eurock 2004 & 53rd Geomechanics Colloquy*, Salzburg, Austria, 61–64.
Gaich, A., Pötsch, M. & Schubert, W. 2006. Acquisition and assessment of geometric rock mass features by true 3D images. In *ARMA Golden Rocks 2006—50 Years of Rock Mechanics*, Golden, Colorado, 17–21 June 2006, Paper 06-1051.
Gaich, A., Pötsch, M. & Schubert, W. 2008. Geologic Mapping with Scaled 3D Images. *The Professional Geologist* 45(6): 48–52.
Haneberg, W.C. 2006. 3-D Rock Mass Characterization Using Terrestrial Digital Photogrammetry. *AEG News 49(4):* 12–15.
Hartley, R. & Zisserman, A. 2000. *Multiple View Geometry in computer vision.* Cambridge: Cambridge University Press.
Heck, J.H. 2004. *Blasting Method.* US Patent No. 6,772,105 B1.
Kilic, A.M., Yasar, E., Erdogan, Y. & Ranjith, P.G. 2009. Influence of rock mass properties on blasting efficiency. *Scientific Research and Essay* 4(11), 1213–1224.
Kim, B.H., Cai, M., Kaiser, P.K. & Yang, H.S. 2007. Estimation of Block Sizes for Rock Masses with Non-persistent Joints. *Rock Mechanics Rock Engineering* 40(2): 169–192.
Mathis, J.I. 1995. A Rock Fragmentation Prediction Model for Mining Excavations. Downloaded from www.zostrich.com.
Moser, P. 2003. Less fines production in aggregate and industrial minerals industry. *Proceedings of the 2nd World conference on explosives & blasting technique*, Prague, Czech Republic, 335–343.
Moser, P., Gaich, A. Grasedieck, A. & Zechmann, E. 2006. The SMX Blast Controller—A new tool to determine the geometrical parameters of a blast based on 3D imaging. In *Proc. of the 32nd Annual Conference on Explosives & Blasting Technique,* Dallas, TX.
Palmström, A. 2001. Measurement and characterization of rock mass jointing. In V.M. Sharma & K.R. Saxena (eds), *In situ characterization of rocks*: 49–97.
Raina, A.K., Ramulu, M., Chourdhury, P.B., Chakraborty, A.K., Sinha, A., Ramesh-Kumar, B. & Fazal, M. 2009. *Productivity improvement in an open-cast coal mine in India using digital image analysis technique.* In J.A. Sanchidrian (ed), *Proc. 9th Int. Symp. on Rock Fragmentation by Blasting—Fragblast 9,* 13–17 September 2009, Granada, Spain, 707–716.
Reichholf, G. 1997. *Untersuchung der Bohr- und Sprengarbeit am Steirischen Erzberg* (German). Master Thesis. University of Leoben. Department of Mining & Tunnelling.
Rogers, S., Moffitt, K. & Chance, A. 2006. Using realistic fracture network models for modelling block stability and ground water flow in rock slopes. In *Sea to Sky Geotechnique 2006*, Vancouver, Canada, 1452–1459.
Roberts, G. & Poropat, G. 2000. Highwall joint mapping in 3D at the Moura mine using SIROJOINT. *Bowen Basin Symposium 2000 Coal and Mining The New Millennium,* Rockhampton.
Scoble, M., Lizotte, Y., Paventi, M. & Doucet, C. 1996. Structural control over fragmentation. In J.A. Franklin & T. Katsabanis (eds), *Measurement of blast fragmentation:* 181–191. Rotterdam: Balkema.
Segarra, P., Sanchidrian, J.A., Moser, P., Grasedieck, A., Smöch, B., & Pascual, J.A. 2005. NOTQUOTEDEnergy controlled blasting for less fines production. *Proceedings of the conference "3rd World conference on explosives & blasting technique",* Brighton, England, 535–543.
Slama, Ch.C. (ed). 1980. *Manual of Photogrammetry.* 4th edition. Falls Church, VA: American Society of Photogrammetry.
Thuro, K. 2002. *Geologisch-felsmechanische Grundlagen der Gebirgslösung im Tunnelbau.* Habilitationsschrift, Münchner Geologische Hefte Reihe B, Heft 18, in German.
Tonon, F. & Kottentstette, J.T. 2007. Laser and Photogrammetric Methods for Rock face Characterization. *Workshop report. 41st U.S. Rock Mech. Symp.—Golden Rocks 2006*, Golden, Colorado.
Wang, L.G., Yamashita, S., Sugimoto, F., Pan, C. & Tan, G. 2003. A Methodology for predicting the in situ size and shape distribution of rock blocks. *Rock Mechanics Rock Engineering* 36(2): 121–142.

Fragmentation testing

Measurement and Analysis of Blast Fragmentation – Sanchidrián & Singh (Eds)
 ISBN 978-0-415-62140-3

Experimental blast fragmentation research in model-scale bench blasts

P. Schimek, F. Ouchterlony & P. Moser
Chair of Mining Engineering, Montanuniversitaet Leoben, Austria

ABSTRACT: This paper summarizes model-scale bench blasts, which were done in 2011 at the Erzberg blasting site of the Montanuniversitaet Leoben. The aim of the project was to compare the fragmentation and the crack development of shots with finite delay (shots-in-a-row) to shots with infinite delay (single-hole-shots in one row), both in virgin and cracked material. The arrangement of the tests in the blasting site ensured that the blasting waves were transmitted to the surroundings. The analysis of the fragmentation was done by sieving and the evaluation of the sieving curves and parameters. The verification of the crack development, which is still in progress, was done by taking thin sections out of drill-cores from the interesting areas after blasting.

1 INTRODUCTION

During the last few years several small scale blasting tests have been done at the Chair of Mining Engineering (e.g. Moser et al. 2003, Reichholf 2003, Grasedieck 2006 and Wimmer 2007). These tests, which were carried out as cylinder-shots or cube-shots with different types of rocks and specimen sizes mainly were done to study the fragmentation or the crack development. The common factor of all these tests was that the specimen was freestanding and therefore they were not comparable with bench-blasts as the blasting waves weren't transmitted to the surrounding rock.

In order to get comparable results with ordinary bench-blasts this project was initiated.

The first part of the project was a comparison of the fragmentation of shots in a sequence (shots-in-a-row with finite delay) to shots with infinite delay (single-hole-shots in one row) with regard to the collective breakage effect. Both types of shots were done in virgin material (first row) and in already damaged material (second row). The fragmentation was verified by doing a sieving analysis of the blasted material and the determination of sieving parameters and sieving curves. The Swebrec distribution (Ouchterlony 2005, 2009) was used as a fitting function because it gives better fits than most other functions (Sanchidrián et al. 2009, 2012).

The second part was the comparison of the cracks, which were caused by the blasting. This was done by surface observations and taking drill-cores of the interesting parts after blasting of the blocks. Out of these drill-cores thin sections were produced and analysed with regard to the development of radial cracks.

2 TEST SET UP

The blasting site of the Chair of Mining Engineering at the Styrian Erzberg was developed during a master thesis (Maierhofer 2011). After the first blasting tests with 2 m³ concrete blocks it was decided to use smaller blocks for further tests. Therefore a yoke inside the walls of the blasting site was used (see Fig. 1).

The gap between the yoke and the walls was filled with compacted sand, which transmitted about 70% of the blasting waves into the surrounding rock. The yoke also has a space where the smaller testing blocks fit into. These small testing blocks, which are made of magnetic mortar (magnetite concrete), are placed on a mat cut out from a used conveyor belt. At the sides and at the back the block is grouted into the yoke by using fast hardening cement, which has the similar material properties as the blocks.

Figure 1. Yoke within the walls of the blasting site that allow waves to escape from test specimen. During the tests the area within the wire fence is covered with rubber mats and heavy non-woven felt to trap the blast fragments.

The basic ingredients and proportions of the magnetite concrete can be seen in Table 1. The recipe had the same proportions as used for small-scale tests at the Luleå Univ. Techn. (Johansson 2008) but used magnetite from Ferroxon instead of Minelco.

The magnetic concrete was produced by a precast concrete plant because they were able to produce large batches of the concrete, about 680 kg. Therefore the produced samples within each batch should have the same properties. To verify the regularity and comparability of the different batches several cylinders with a diameter of 138.5 mm and a height of 280 mm have been produced from each batch. These cylinders were blasted with a 20 g/m detonating cord in a 10 mm diameter blasthole and a sieving analysis was done as a measure of the repeatability of the fragmentation properties.

2.1 *Testing blocks*

The dimensions of the testing blocks which were used for blasting were 660 × 280 × 210 mm (L×H×W) – the same as Johansson & Ouchterlony (2012) used.

The blastholes with a diameter of 10 mm were drilled in the laboratory using core-drilling equipment. The advantage of this equipment was that the spalling effect at the bottom of the blocks, which occurred during ordinary hammer drilling, was minimized and the deviation of the blastholes was within a few millimetres.

2.2 *Shots with finite delay*

The shots with finite delay (see Table 2) were carried out as single-row-shots (2 rows per testing block) with a burden of 70 mm for each row. The side-spacing of the blastholes was 110 mm (5 holes per row) respectively 95 mm (7 holes per row). This arrangement gives a ratio of side-spacing to burden (S/B) of 1.57 and 1.36.

The delay between the holes was decided to be 2 milliseconds per meter (ms/m) of burden. Due to the geometrical properties of the blocks (width of 210 mm) and the chosen 2-rows of blastholes the delay between the holes has to be 140 μs. This delay was achieved by using a 5 g/m detonating cord which was lying in loops on a piece of a conveyor belt. This arrangement was used to protect the yoke from the detonation of the delay-timing cord.

The blasting was done by using a 20 g/m detonating cord. The finite delay shots with 5 holes per row started from the left side in the 1st row and from the right side in the 2nd row, as seen from the front, while the blasting of the 7 holes per row was initiated in the middle of the row.

Table 1. Ingredients of magnetite concrete blocks.

Ingredient	[%]
Portland cement CEM II/A-M 42.5 N	25.60
Water	12.65
Glenium 361 (Plasticizer)	0.256
DCC-Entlüfter (Defoamer)	0.129
Magnetite powder (Ferroxon 618)	29.65
Quartz sand 0.1–0.5 mm (ME 31)	31.70

Table 2. Overview of the shots with finite delay.

Block #	Row #	Holes per row	Type of initiation
1	1	5	Left to right
1	2	5	Right to left
2	1	7	Middle of the row
2	2	7	Middle of the row

2.3 *Shots with infinite delay*

It was decided to use the arrangement with 5 holes per row for the shots with infinite delay (single-hole-shots). The side-spacing (110 mm) and the burden (70 mm) were the same as described above.

During several single-hole-shots in the middle of the rows the influence of blasting to the neighbouring holes of the row was investigated. These investigations were done with different filling materials of the neighbouring holes of the row. One test was carried out with unfilled holes, one with wooden sticks in the holes, one with fast hardening cement in the holes and the last one with undrilled neighbouring holes. All of these tests were carried out as single-hole-shots in the middle of the row. Table 3 gives an overview over the single-hole-shots in the middle of the rows.

Additionally to these tests two testing blocks were blasted with single-hole-shots, which started from one side of the block (see Table 4).

Table 3. Overview of the single-hole-shots in the middle of the rows.

Block #	Row #	Filling material of neighbouring holes
3	1	Wooden sticks
3	2	No filling material
4	1	Fast hardening cement on the right side; No holes drilled on the left side
4	2	Fast hardening cement on the right side; No holes drilled on the left side

Table 4. Overview of the single-hole-shots which started from one side of the block.

Block #	Row #	Holes per row	Filling material
5	1	5	Fast hardening cement
5	2	1	No filling material
6	1	5	No filling material
6	2	5	No filling material

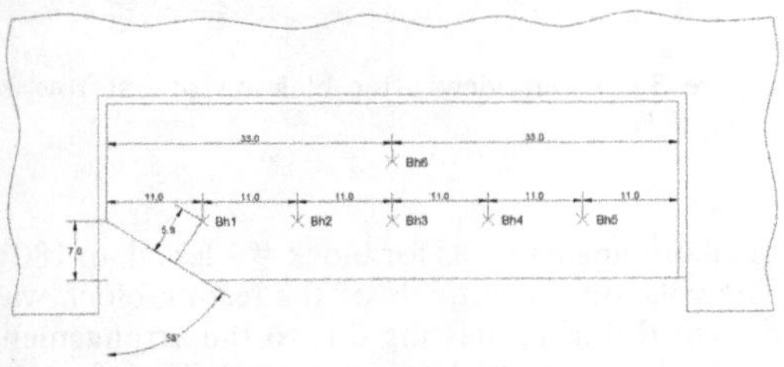

Figure 2. Geometrical properties of Block # 5.

This ensures same geometrical properties for the following holes of the rows.

In order to simplify the throwing of the burden of the first single-hole-shot it was decided to cut away one corner of the block. This would guarantee a smaller burden for the first blasthole. Figure 2 shows the geometrical properties of this "cut-away-corner" arrangement in block # 5.

2.4 *Sieving analysis*

The blasted material was collected after each blast and a sieving analysis was done. As a result of the sieving analysis the particle size distribution was created. Out of this the local inclinations as well as several k-values were calculated.

The blasted material was sieved in two steps: The grains of the coarser material were analysed individually by sticking them through the mesh of the sieves. The following sieves were used for that procedure: 125; 100; 80; 63; 50; 40; 31.5; 25; 20; 14; 12.5; 10 mm.

The finer material was sieved by hand using the following screen sizes: 6.3; 4; 2; 1; 0.5; 0.25 mm. It was decided not to use a sieving machine because secondary breakage while sieving should be avoided. After the first few sieving analyses it was decided to stop the screening at a grain size of 0.5 mm because the ingredients of the magnetic mortar (see Table 1) have grain sizes up to this size.

During each sieving step an optical sorting process was carried out to minimize contamination. This was done because the lighter pieces came from the fast hardening cement or the yoke and the dark pieces were particles from the test block (magnetic mortar).

The result of the sieving analyses was plotted in a GGS-diagram (Schuhmann 1940). This diagram is a log-log-plot where the x-axis shows the screen size and the y-axis shows the cumulated mass passing.

In addition to the particle size distribution several k-values were calculated. The specific k-value was calculated with linear interpolation between the two grain sizes next to the cumulative mass passing at 30%, 50% and 80%.

2.5 *Analysis of crack development*

The investigations of differences between the crack development of shots with finite delay (shots-in-a-row) and shots with infinite delay (single-hole-shots) have been done by taking drill-cores of the interesting areas after blasting. Therefore the gap in the yoke after blasting the testing block was filled with fast hardening cement. Drill-cores were taken directly behind blasted holes, between blasted holes and around blasted holes.

The drill-cores were used for several Brazilian tests as well as for production of thin sections.

The results of these tests are still under investigation.

3 RESULTS

3.1 *Material properties*

Several investigations with regard to the material properties of the used magnetite concrete have been done by Khormali (2011). Table 5 shows the results.

Out of each produced batch of magnetite concrete several cylinder have been produced as described in section 2. Table 6 gives an overview of the sieving parameters of the cylinder shots (one cylinder per batch was blasted).

Table 5. Material properties of the used magnetite concrete (Khormali 2011).

Material property	Mean	Std. dev.	Unit
Density	2.38	0.01	[g/cm³]
UCS	62.7	2.1	[MPa]
Brazilian tensile strength	4.98	0.52	[MPa]
Young's modulus	25.0	0.4	[GPa]
Poisson's ratio	0.17	0.01	[–]
P-wave velocity	3865	40	[m/s]
S-wave velocity	2330	29	[m/s]

Table 6. Overview over the sieving parameters for the cylinders of the different batches.

Batch #	k_{30} (mm)	k_{50} (mm)	k_{80} (mm)
1	6.1	11.4	19.3
2	6.7	12.2	20.6
3	6.0	11.0	18.3
4	6.2	11.2	19.1

The sieving parameters show reasonably comparable results for the different batches. Compared to former cylinder-shots with similar ingredients the k_{50} is lower. Johansson (2008) blasted 15 cylinders with the same properties and similar ingredients and got an average k_{50} of 15.3 mm with a standard deviation of 1.1 mm.

3.2 *Shots with finite delay*

As described in section 2.2, two testing blocks, one (block # 1) with 5 holes per row and a second (block # 2) with 7 holes per row were shot with a delay of 140 μs.

3.2.1 *Blasting results*

The result of blasting the 1st row with 5 holes in block # 1, which can be seen in Figure 3, was that each of the five half-casts was visible after the blasting. The breakout angle between the individual holes was approximately 180° but at the end of the block the angle was smaller. A crack to the sides of the block was also developed in this area, see left and right side in Figure 3. At the top of every blasthole a thin section broke off. A similar phenomenon was observed by Reichholf (2003) and Wimmer (2007). Between the 2nd and the 5th blasthole the meeting points, or lines rather, of the cracks were clearly visible as a trench.

When blasting the second row with 5 holes in block # 1 the detonating cord used for the delay-timing was cut before the detonation reached the 5th blasthole. Therefore just the charges in blastholes 1 to 4 detonated. The breakout angle between the 4 holes was approximately 180°. To the side of the testing block the breakout angle was less than 180° but in the 2nd row the crack to the side of the block (which was detected in the 1st row) was not detected. At the top of the blastholes thin sections broke off and between the 2nd and 3rd blasthole there was a significant backbreak.

After blasting of the 1st row with 7 holes in block # 2 (initiation in the middle of the row) all of the half-casts were again visible. The breakout angle between the seven blastholes was approximately 180° but at the side of the block the breakout angle was as for block # 1 less than 180°. The yoke on the left side of the testing block was damaged during blasting due to the arrangement of the delay-timing detonating cord. Therefore the breakout angle at this side of the block was also nearly 180°. At the top of the blastholes thin parts broke off.

Figure 3. Front view after blasting of 1st row in block # 1.

After blasting of the 2nd row with 7 holes in block # 2 also all of the half-casts were visible. The breakout angle between the seven blastholes was approximately 180° and at the side of the block the breakout angle was less than 180°. Due to the fact that the yoke on the left side of the testing block was already damaged during blasting the breakout angle at this side of the block was also nearly 180°.

3.2.2 *Speed of crack development*

If the speed of the crack-development is lower than 790 m/s (side-spacing of 110 mm divided by a delay of 140 μs) the meeting point of the cracks, which are generated by blasting should lie somewhere in between the blastholes. For the calculation of the actual speed of the crack development the judged meeting point is determined on the taken pictures after blasting. The distance between the meeting point and the later initiated blasthole is also determined. The same distance is measured from the meeting point to the direction of the first initiated blasthole. The cracks of the first hole have developed until this point when the next hole is initiated. With this distance the material-specific speed of crack-development can be calculated. The average speed of crack-development in the used magnetite concrete was determined to be ~260 m/s. This is quite low compared to Johansson & Ouchterlony (2012) who determined ridges on the surface of a blasted magnetite block as the meeting point of cracks and calculated the speed of crack-development to be about 650 m/s.

3.2.3 *Fragmentation results*

The sieving parameters are given in Table 7 that shows k_{50}-values [mm] for row 1 and row 2 when blasting 5 and 7 holes-per-row.

It can be seen, that in both cases (5 and 7 holes per row) the sieving parameter k_{50} of the second row is less than k_{50} for the first row. This fact was also seen by Johansson & Ouchterlony (2012).

The k_{50}-value of row 2 for the 5 holes-in-a-row shots is 42.2% less than the k_{50}-value of row 1 while the k_{50}-value of row 2 for the 7 holes-in-a-row shots is 31.3% less than the k_{50}-value of row 1.

Figures 4 to 6 show the results of the sieving analysis with the Swebrec function used for curve fitting. It can be seen that the results of the finest sieve used (0.25 mm) aren't well reproduced by the basic three parameter Swebrec function. Using the 5-parameter Swebrec function would 'bend' the curve downward and improve this.

Table 7. k_{50}-values for shots with finite delay.

	Row 1 (mm)	Row 2 (mm)
5 holes per row (Block # 1)	28.9	16.7
7 holes per row (Block # 2)	17.6	12.1

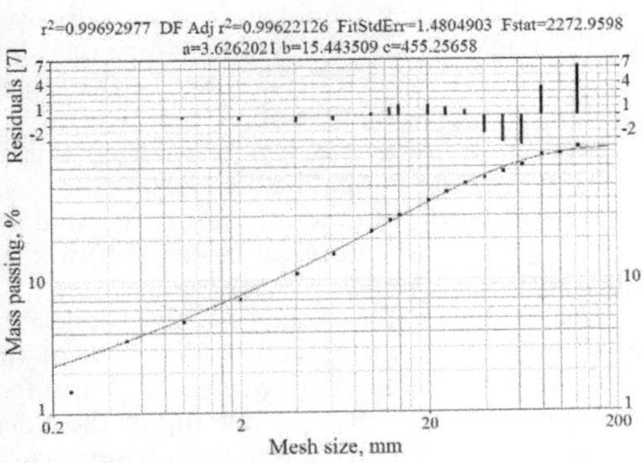

Figure 4. Sieving curve for row 1 of block # 1.

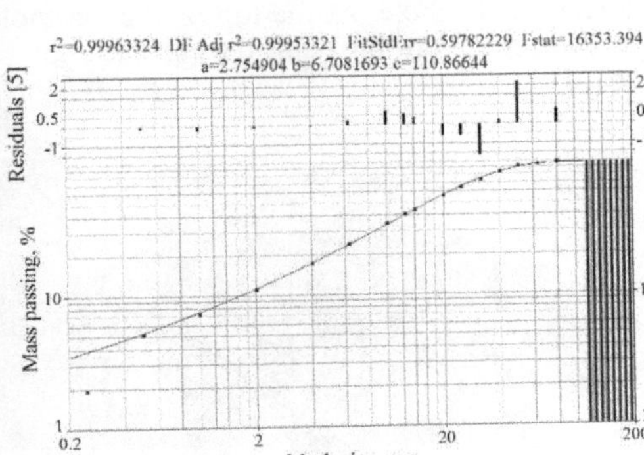

Figure 5. Sieving curve for row 2 of block # 1.

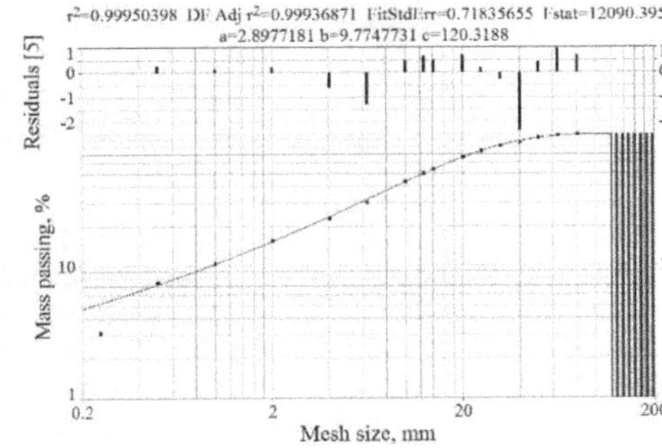

Figure 6. Sieving curve for row 2 of block # 2.

When the results of the second row are compared to the results of the first row, it can be seen that the fragmentation of the second row is much more even than the fragmentation of the first row (evaluating the residuals). This result is most likely influenced by the fact that the first row contains virgin material and the second row contains already cracked material (see also Johansson & Ouchterlony 2012).

3.3 *Shots with infinite delay*

As described in chapter 2.3 two testing blocks (# 3–4) were blasted with single-hole-shots in the middle of the rows (2 rows per testing block) and two blocks (# 5–6) were shot with the cut-away corner arrangement.

3.3.1 *Blasting results*

3.3.1.1 Single-hole-shots in the middle of the row

When filling the neighbouring holes of the rows with wooden sticks (the diameter of these sticks was slightly smaller than the diameter of the blast-holes) the neighbouring holes were completely split open after blasting and a dominant crack has developed through them until the end of the block. For some reasons the movement of the burden at the right side was much larger than on the left side.

When the neighbouring holes weren't filled with any material (blasting of the 2nd row) these holes were also destroyed by blasting but the dominant crack through the other holes until the end of the block was not developed.

Due to suggested different reflections of the blasting waves the holes on the right side of the next testing block were drilled in the laboratory and filled with fast hardening cement while on the left side just a shallow collar of about 5 mm depth was drilled. After blasting the neighbouring holes, especially at the right side of the block, were destroyed and a dominant crack through the entire block has developed. For some reasons the crack at the right side developed slightly to the back while on the left side the crack developed parallel to the free surface.

When blasting the hole in the middle of the second row with the same arrangement the neighbouring blastholes are also heavily influenced by blasting but the dominant crack through the entire block couldn't be detected.

3.3.1.2 Single-hole-shots with a cut-away corner

The first testing block with this arrangement (block # 5) was prepared with 5 blastholes in the first row and 1 blasthole in the middle of the second row. The holes were filled with fast hardening cement and drilled just right before blasting. The blocks were painted and then repainted between each shot to better visualize the breakage.

After blasting the first hole (see Fig. 7) the breakout at the top of the block was larger than at the bottom and a thin section around the top of the blasthole broke off. Furthermore a dominant crack, which cut through the neighbouring holes was developed. This crack goes directly through the second blasthole, then it travels between the second and third blasthole to the back side of the testing block with a sharp change in direction almost in the middle between the two holes. After a further kink between the 3rd and 4th blasthole this crack travelled to the front side of the block. A similar crack development was detected by Petropoulos (2012) when a row of holes with large initiation delay of the charges was blasted.

To be able to blast the second hole it was necessary to drill the hole with a percussion-drilling-machine. Due to the vibrations generated by drilling, the crack, which has developed during blasting of the first hole, was activated and the burden for the second hole has been loosened. Therefore it was just put back in place there right before the blasting was done. Due to this fact just a few new cracks at the bottom of the burden have been developed during blasting.

After drilling of the third hole the detonating cord at the top of the blasthole had no confinement due to breakout while drilling. Therefore just the burden at the bottom of the block was thrown, but some small cracks have developed.

While drilling the 4th hole the small cracks which have been developed during blasting were activated and the burden was loosened. Therefore it was just put back in place before blasting. Additionally to this loosened burden a small part at the bottom of the block was thrown during blasting.

After the 5th blasting the breakout-angle between the blasthole and the end of the block was less than 180° but otherwise similar to the angle which occurred during the shots-in-a-row. The dominant crack to the end of the block (which occurred during the shots-in-a-row) couldn't be detected.

The result of the single-hole-shot in the middle of the second row was similar to the results seen in section 3.3.1.1. If there would have been neighbouring holes they would have been destroyed by blasting (see Fig. 8). A dominant crack has developed just to one direction and at the top of the block a thin slab broke off.

Due to the problems which occurred during drilling of the holes right before blasting the holes of the next block with the cut-away corner and the 5 holes per row arrangement (block # 6) weren't filled with any material.

The blasting result of the first single-hole-shot (see Fig. 9) was quite similar to the result of the first block with the "cut-away-corner" arrangement. The larger breakout at the top of the block and even the dominant crack at the top were nearly the same. For some reasons this crack also travelled to the back of the block between the 2nd and the 3rd blasthole but changed direction right before passing the 3rd hole. At the top of the blasthole also some thin slabs broke off during blasting.

Figure 7. Block # 5 after blasting of the 1st blasthole. Note bend in crack in blue section between holes # 2–3 and the kink between the holes # 3–4 in grey section.

Figure 8. Front view after blasting of a single hole in the middle of row 2 of block # 5.

Figure 9. Block # 6 after blasting of the 1st blasthole.

As a result of the second single-hole-blast the crack which was developed during the first shot was activated and this material was thrown out. Also some more cracks at the top and at the front of the block were developed.

Due to the enormous breakout of the second shot nearly half of the detonating cord placed in the third blasthole detonated on the free surface. The already existing cracks were activated during the shot and the fourth blasthole was completely destroyed. One new crack at the top of the block was generated. This crack grew through the fifth hole in the second row until the side of the block.

Due to the enormous breakout of the third single-hole-shot it was impossible to blast the fourth hole. Therefore the next single-hole-shot was done in the fifth blasthole. The breakout-angle between the blasted hole and the end of the block was smaller than 180° but similar than the breakout-angle which was seen in block # 5.

In order to get comparable results from the blasting of the first and the second row the corner of the second row was also cut-away before blasting.

As a result of blasting the first hole the second hole was destroyed and therefore no blasting in this hole could be done.

Due to this excessive damage the second single-hole-shot in the second row was carried out in the third hole. As a result of blasting a crack which goes from the blasthole to the back of the block was generated. Also a large part of the burden of the 4th hole was already thrown during this shot.

During blasting of the next hole a connecting crack to the crack which was generated while blasting the 3rd blasthole in the first row was developed. Also several small cracks at the top of the block were developed and a thin part at the top of the block broke apart.

The blasting of the last blasthole in the second row was carried out as planned. Due to the blastings before the burden of this blasthole was already loosened from the rest of the block. Therefore the crack which was developed upfront was activated and this part of the burden was thrown. The breakout-angle between the blasthole and the end of the block was 180° due to the fact that this crack already existed upfront.

3.3.2 *Fragmentation results*

Table 8 shows the results of the single-hole-shots in the middle of the rows while the other holes were filled with fast hardening cement (right side) respectively not drilled before blasting (left side).

It can be seen that the sieving parameter k_{50} of the second row is 12% less than k_{50} of the first row.

The sieving parameter k_{50} for the single-hole-shots, which were carried out with the "cut-away corner" arrangement were evaluated individually for each hole and as a sum for the entire row. This was done by summing up the masses of the individual sieving steps for each blasted hole.

Table 9 shows the results of single-hole-shots with "cut-away corner" arrangement while the neighboring holes were filled with fast hardening cement. In the second row just one blasthole in the middle of the row was shot.

As a result of the sieving parameters it can be seen that the variation of k_{50} of the individual blastholes in the first row is quite large. The finest fragmentation was achieved at the 5th blasthole where the k_{50} is 64.8% less than the k_{50} for the 2nd blasthole.

The comparison of the sum of the k_{50} values for the individual blastholes in the 1st row and the k_{50} for the blasthole in the 2nd row shows that the k_{50} for the second row is 15.9% less than k_{50} for the first row.

Table 10 shows the results of single-hole-shots with "cut-away corner" arrangement while the neighboring holes were not filled. With this

Table 8. k_{50}-values for single-hole-shots in the middle of the rows (Block # 4).

	Bh 1	Bh 2	Bh 3 (mm)	Bh 4	Bh 5
Row 1			55.8		
Row 2			49.1		

Table 9. k_{50}-values for single-hole-shots with "cut-away corner" arrangement and filled neighboring holes (Block # 5).

	Bh 1 (mm)	Bh 2 (mm)	Bh 3 (mm)	Bh 4 (mm)	Bh 5 (mm)	Sum 1–5 (mm)
Row 1	19.8	48.6	20.4	38.3	17.1	25.8
Row 2			21.7			

Table 10. k_{50}-values for single-hole-shots with "cut-away corner" arrangement and not filled neighboring holes (Block # 6).

	Bh 1 (mm)	Bh 2 (mm)	Bh 3 (mm)	Bh 4 (mm)	Sum 1–4 (mm)
Row 1	17.2	30.5	94.3	20.1	28.1
Row 2	16.9	23.6	12.9	25.4	20.6

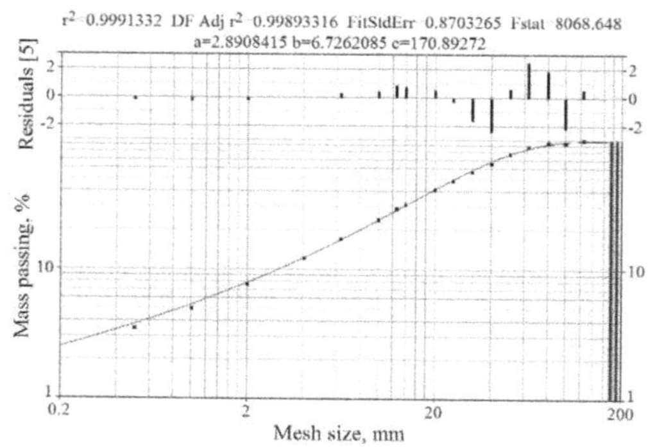

Figure 10. Sieving curve for the sum of the single hole shots in row 1 of block # 5.

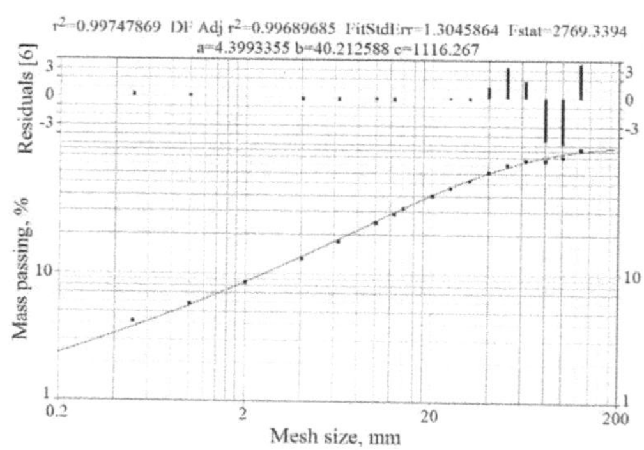

Figure 11. Sieving curve for the sum of the single hole shots in row 1 of block # 6.

arrangement just 4 blastholes in each row has been blasted due to the extensive breakout of some single-hole-shots. In row # 1, borehole 4 couldn't be shot, in row # 2 borehole 2 couldn't be shot.

Also with this arrangement the huge variation of the k_{50} values for the individual shots can be seen, especially in the first row where the smallest value is 81.7% less than the highest value. The smallest value in the second row is 49.2% less than the highest value in this row.

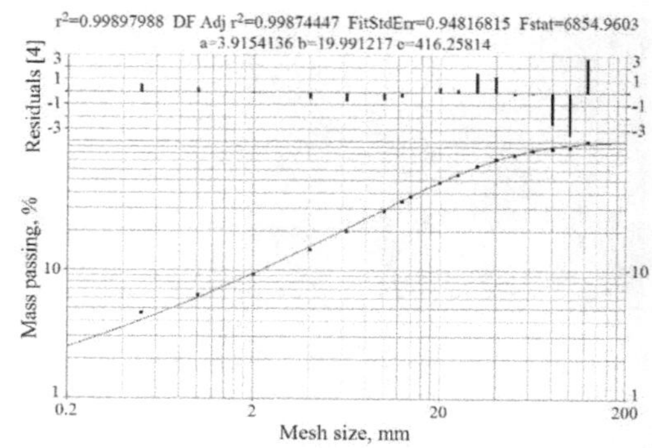

Figure 12. Sieving curve for the sum of the single hole shots in row 2 of block # 6.

The comparison of the sum of the k_{50} values for the individual blastholes in the 1st row and the 2nd row shows that the k_{50} for the second row is 26.7% smaller than k_{50} for the first row.

Figures 10 to 12 show the results of the sieving analysis with the Swebrec function used for curve fitting. The figures show the sum of the single hole shots in the individual rows.

It can be seen that the Swebrec function describes the distribution of the blasted material quite well.

4 DISCUSSION

4.1 *Blasting results*

If a delay of 2 milliseconds per meter of burden is chosen for shots-in-a-row then the meeting point of the cracks is between the blastholes. The meeting point of the dominant cracks was clearly visible as a trench after blasting the 1st row of the 5 shots-in-a-row arrangement. The calculated speed of the cracks of about 260 m/s seems to be low compared to figures in the literature: Wilson & Holloway (1987) determined the speed of radial cracks in concrete models between 1000 m/s (near the boreholes) slowing down to around 200 m/s far away from the boreholes. Also Johansson & Ouchterlony (2012) found a higher speed of crack-development of about 650 m/s.

An initiation in the middle of the row works for shots in the same sequence, at least with our chosen delay, but doesn't work for holes shot one by one. In the case of single-hole-shots the neighbouring holes are heavily damaged by the blasting. For single-hole-shots in the middle of the 2nd row it doesn't matter if in the 1st row just one hole or several holes have been shot. The result in each case is that the single-hole-shot destroys the neighbouring holes.

Blasting of single-holes showed a breakout angle which is larger than the often in textbooks assumed 90° (Johansson & Persson 1970). This is the case for shots in the middle of the row as well as for shots at one corner of the block (with a cut-away corner).

At the top of nearly every blasted hole a thin slab behind the blasted hole broke off. Reichholf (2003) and Wimmer (2007) also observed such a phenomenon. Wimmer (2007) attributed the reflection of tensile waves at the free surface to be the reason for these breakouts. Another reason for this behaviour might be the brittle properties of the material which were encountered during tests with percussive-drilling equipment. In these tests spalling at the bottom of the testing blocks was observed. Another reason for this behaviour might be the fact that the 20 g/m detonating cord, which was used for blasting, stood out a few centimetres of the blastholes in order to make sure that the delay-timing with the 5 g/m detonating cord works properly.

After single-hole-shots in the middle of the 1st row a significant crack parallel to the free surface was developed. This crack was detected in the case when there was no filling material in the neighbouring holes, in the case when the neighbouring holes were filled with fast hardening cement and also in the case when there were no neighbouring holes. Therefore this crack was not influenced or attracted by the properties of the neighbouring holes. The crack was not that significant when blasting in the middle of the 2nd row.

The breakout of single-hole-shots in the middle of the row was not symmetrical but larger at the right side of the blocks. The reason for this effect might be that the detonating cord was not centralized in the blasthole and therefore not decoupled from the walls (especially at the top of the blasthole).

The asymmetric breakout in the case of the 1st shot with the "cut-away-corner" arrangement, which showed a larger breakout at the top might be also influenced by the fact of irregular coupling of the detonating cord.

The 1st single-hole-shot in the 1st row with the "cut-away-corner" arrangement developed a significant crack, which started parallel to the free surface. After passing the 2nd hole it changed the direction to the back of the testing block. Right before the 3rd hole the crack made a nearly 90° turn and travelled through the 3rd hole to the front of the block. The reason for this sudden change in direction might be found in the blasting waves. This significant crack with the sharp turn was detected with filled neighbouring holes as well as with unfilled neighbouring holes. Therefore the assumption that the crack was not influenced or attracted by the properties of the neighbouring holes can be made.

4.2 *Fragmentation results*

Blasting of 7 holes per row gives a finer fragmentation than blasting of 5 holes per row when the size of the specimen and the burden of the individual rows are the same. Therefore the specific charge for the broken volume of rock is higher in the 7 holes-per-row arrangement.

If the blasts in the 1st row are done in virgin material then the fragmentation in the 2nd row is always finer than in the 1st row (see also Johansson & Ouchterlony, 2012). The reason for this observation is probably the pre-conditioning of the burden for the 2nd row while blasting the 1st row. This observation was made for shots with finite delay as well as for shots with infinite delay.

The fragmentation of blasts in the 2nd row is heavily influenced by the number of blasted holes in the 1st row. The k_{50}-value for a single-hole-shot in the middle of the 2nd row is 55.8% less if in the 1st row 5 single-hole-shots are carried out compared to only one single-hole-shot in the middle of the 1st row. This observation is also linked to the pre-conditioning of the burden for the 2nd row by blasting the 1st row.

There is just a minimal effect on the fragmentation comparing holes shot in the same sequence and the sum of holes shot one by one. This is the case even if one hole out of five couldn't be shot in some cases of the single-hole shots.

There is a huge variation between the k_{50} -values of the individual single-hole-shots due to the fact that the neighbouring holes are quite heavily influenced by the previously shot blastholes.

5 CONCLUSION

Summing up the results of the investigations, which have been done with several arrangements, the following conclusions can be drawn:

- An initiation in the middle of the row works for shots in the same sequence but doesn't work for holes shot one by one due to the damage of the neighbouring holes of the row.
- Blasting of single-holes showed a breakout angle which is larger than 90°.
- Blasting of 7 holes per row gives a finer fragmentation than blasting of 5 holes per row (same size of specimen).
- If the blasts in the 1st row are done in virgin material then the fragmentation in the 2nd row is always finer than in the 1st row.

- The fragmentation of blasts in the 2nd row is heavily influenced by the number of blasted holes in the 1st row.
- There is just a minimal effect on the fragmentation comparing holes shot in the same sequence and the sum of holes shot one by one.

Detailed results of the investigations, including the thin sections with cracking and surface properties of the bench face, will be found in Schimek (2012).

ACKNOWLEDGEMENT

The authors wish to acknowledge the work at Swebrec (Swedish Blasting Research Centre at Luleå University of Technology) where the testing with the present set up started. We also wish to acknowledge the work of Erhard Maierhofer with regard to the documentation of the blasting results. Finally, we would like to thank Gerold Wölfler and Arno Hofmann for their support during the blasting tests.

REFERENCES

Grasedieck, A. 2006. Die natürliche Bruchcharakteristik (NBC) von Gesteinen in der Sprengtechnik. Doctoral Thesis, Montanuniversitaet Leoben, Chair of Mining Engineering and Mineral Economics.

Johansson, C.H. & Persson, P.A. 1970. *Detonics of High Explosives*. London: Academic Press.

Johansson, D. 2008. Fragmentation and waste rock compaction in small-scale confined blasting. Licentiate thesis, Lulea University of Technology, Division of Mining and Geotechnical Engineering.

Johansson, D. & Ouchterlony, F. 2012. Shock wave interactions in rock blasting—the use of short delays to improve fragmentation in model-scale. *Rock Mechs & Rock Engng*. In press.

Khormali, R. 2011. An investigation into the natural breakage characteristics (NBC) of rock and rock-like materials in blasting, and model correlations in fragmentation phenomena. Manuscript under preparation, Montanuniversitaet Leoben, Chair of Mining Engineering and Mineral Economics.

Maierhofer, E. 2011. Development of a blasting area and blasting tests with concrete blocks in half scale to research different fragmentation phenomena. Master thesis, Montanuniversitaet Leoben, Chair of Mining Engineering and Mineral Economics.

Moser, P., Olsson, M., Ouchterlony, F. & Grasedieck, A. 2003. Comparison of the blast fragmentation from lab-scale and full-scale tests at Bårarp. In R. Holmberg (ed.), *Proc. EFEE 2nd World Conf. on Explosives & Blasting Techn*, pp 449–458. Rotterdam: Balkema.

Ouchterlony, F. 2005. The Swebrec© function, linking fragmentation by blasting and crushing. *Mining Technology (Trans. Inst. Min. Metal A)* 114: A29–A44.

Ouchterlony, F. 2009. Fragmentation characterization; the Swebrec function and its use in blast engineering. In J.A. Sanchidrián (ed.), *Proc. 9th Int. Symp. on Rock Fragmentation by Blasting—Fragblast 9*. Leiden: CRC Press, pp. 3–22.

Petropoulos, N., Johansson, D. & Ouchterlony, F. 2012. Fragmentation under different confinement conditions and the burden behavior-small scale tests. *Fragblast 10 Workshop on Fragmentation measurement and analysis, New Delhi, India, 24–25 November 2012*.

Reichholf, G. 2003. Experimental investigation into the characteristic of particle size distributions of blasted material. Doctoral Thesis, Montanuniversitaet Leoben, Chair of Mining Engineering and Mineral Economics.

Sanchidrián, J.A., Segarra, P., López, L.M. & Ouchterlony, F. 2009. Evaluation of some distribution functions for describing rock fragmentation data. In J.A. Sanchidrián (ed.) *Proc. 9th Int. Symp. Rock Fragmentation by Blasting*. Leiden: CRCPress/Balkema, pp. 239–248.

Sanchidrián, J.A., Ouchterlony, F., Moser, P., Segarra, P. & López, L. 2012. Performance of some distributions to describe rock fragmentation data. *Int. J. Rock Mech. Min. Sci.* 53: 18–31.

Schimek, P. 2012. Experimental blast fragmentation research in model-scale bench blasts. Research report under preparation. Montanuniversitaet Leoben, Chair of Mining Engineering and Mineral Economics.

Schuhmann, R. 1940. Principles of comminution, 1-Size distribution and surface calculation. *Tech. Publs. AIME*, no. 1189,11.

Wilson, W.H. & Holloway, D.C. 1987. Fragmentation Studies in instrumented concrete models. In G. Herget & S. Vongpaisal (eds.), *Proc. 6th ISRM International Congress on Rock Mechanics, Volume 1*. Rotterdam: Balkema, pp. 735–741.

Wimmer, M. 2007. An experimental investigation of blastability. Swebrec rpt 2007:1. Lulea: Swedish Blasting Research Centre at Lulea Univ. Techn.

Measurement and Analysis of Blast Fragmentation – Sanchidrián & Singh (Eds)
© 2013 Taylor & Francis Group, London, ISBN 978-0-415-62140-3

Fragmentation under different confinement conditions and the burden behaviour—small scale tests

N. Petropoulos
Luleå University of Technology, Division of Mining and Geotechnical Engineering, Sweden

D. Johansson
Swebrec, Swedish Blasting Research Centre at Luleå University of Technology, Sweden

F. Ouchterlony
Swebrec, Swedish Blasting Research Centre at Luleå University of Technology, Sweden
Montanuniversität Leoben, Austria

ABSTRACT: Small scale-tests have been performed on magnetic mortar blocks to investigate the influence of delay times, specific charge and firing pattern on fragmentation both when the burden is free and when confined by debris simulating the confined conditions of a SLC-blast ring and to study the burden behavior during blasting by implementing dynamic methods of measurements. The dimensions of the test blocks were 660 × 270 × 210 mm (L × W × H). Two different blast hole patterns, 2 rows with 5 holes with a burden of 70 mm (S/B = 1.6) and 3 rows with 7 holes with a burden of 58.3 mm (S/B = 1.4). A complete analysis regarding the influence of different delay times ranging from 0 to 4.1 ms/m of burden, different specific charges and different firing patterns on fragmentation has been made, based upon 22 blocks.

1 INTRODUCTION

Sub-Level Caving (SLC) is an important mass mining method used in several mines around the world with different geological conditions. The burden of a SLC-blast ring is confined by caved rock, which has a negative impact on fragmentation (Johansson & Ouchterlony 2011). This study was conducted to investigate any potential improvement of fragmentation by changing key parameters.

Researchers in the blasting field attempt on a daily basis to improve the results of bench or underground blasting. Recent studies focus on the effect of short delay times on fragmentation by means of potential wave interaction.

Stagg & Nutting (1987) conducted 24 small scale blast tests with different delay times from 0 to 118 ms/m of burden. The experimental procedure was to use a 1.14 m dolomite bench with burden distance 0.38 m and two different spacing distances, 0.53 and 0.76 m giving a spacing/burden ratio of 1.4 and 2.0 respectively. The best fragmentation was observed between 36 ms/m and 55 ms/m of burden due to an interaction of stress waves with strains induced by gas pressure of the previously detonated blasthole. A second series of similar blasting tests were conducted by Stagg & Rholl (1987) with a different test scale. The designs of the tests included 6.7 m limestone benches with burden distances from 1.7 to 1.8 m formed by local irregularities and a spacing of 2.1 m resulting in a specific charge of 0.59 kg/m³. The detonation was induced by seismic caps with 1.10 ms ± 0.11 ms timing accuracy. The results showed that there was an improvement in fragmentation for delays between 3.3 to 26 ms/s of burden. They concluded that the optimum fragmentation was obtained when the stress wave interacts with the ongoing failure process of the previous hole.

Katsabanis et al. (2006) made small scale blast tests to investigate the effect of short delays on fragmentation. The tests were performed on a small scale granodiorite block. The drilling pattern composes an equilateral triangular spacing of 102 mm giving a burden of 8.8 cm. The tested delays were 0, 10, 20, 40, 80, 100, 1000, 2000 and 4000 μs. The purpose of the very short delays was to investigate the fragmentation under shock wave interactions. The blocks were free (four free faces) trapping all the produced stress waves within the block body. The findings of these tests were in agreement with previous tests (Stagg & Rholl 1987), indicating the delay time for optimum fragmentation to lay in the range of 3.3–13 ms/m of burden.

Vanbrabant & Escobar (2006) made full scale blasting tests to observe the fragmentation at short delays by means of electronic detonators. Their design was based on the Rossmanith et al. (1997)

procedure. The choice of the delay time was based on the concept of overlapping the negative tail of the P-wave particle velocity. The implementation of this technique gave an improvement in x_{50} by 45.6%.

Johansson & Ouchterlony (2012) blasted 15 small scale magnetic blocks using short delays to investigate the behavior of fragmentation under the condition of shock wave interaction. The magnetic blocks were confined by a U-shaped yoke and their dimensions 660 × 270 × 210 mm (L × W × H). Each block had two rows with five holes each on a pattern B × S = 70 × 110 mm There were two different set-ups of the tests, confined, there is material lying in front of the burden, and free face conditions. The delay time varied from instantaneous (0 µs) to 146 µs. The approach of estimating the delay time was based on Vanbrabant's scheme (Vanbrabant & Escobar 2006). The results showed a plateau for the median fragment size (x_{50}) for free conditions for row #2 in the range of 0.65–1.1 ms/m of burden. Additionally, the mean size of the past blast material was coarser in the confined tests by almost 100%.

2 METHODOLOGY

2.1 *Fragmentation measurements*

The set-up of the tests was inspired by a previous study (Johansson & Ouchterlony 2012). The tests were composed of magnetic mortar blocks. The properties of the magnetic mortar are well-defined by previous studies and are similar to the magnetite orebody in the Kiruna mine (Johansson et al. 2007, 2008, Johansson & Ouchterlony 2011). The magnetic mortar blocks were made of magnetite powder, cement, quartz-sand and water. The purpose of using magnetic mortar blocks was that it behaves in similar way as the magnetite ore in the Kiruna mine. The dimensions of the blocks were 660 × 270 × 215 mm (L × H × W) and the blasthole diameter was ø10 mm. In this study, two designs were tested. The first design had 5 holes and 2 rows, the burden was 70 mm and the spacing distance was 110 mm giving a spacing/burden ratio 1.6. The second design had 7 holes and 3 rows, the burden was 58.3 mm and the spacing was 82.5 mm giving ratio spacing to burden 1.4. The magnetic blocks were confined by a U-shaped yoke to minimize the reflection of waves (Fig. 1).

The yoke was made of high strength concrete. The dimensions of the yoke were 250 mm width and 350 mm high. There was a gap between the yoke the block of around 10 mm. It was filled with fine grained expanding grout (Weber WXM 702), which has the same properties as the concrete in the yoke, minimizing the impedance of the shock waves.

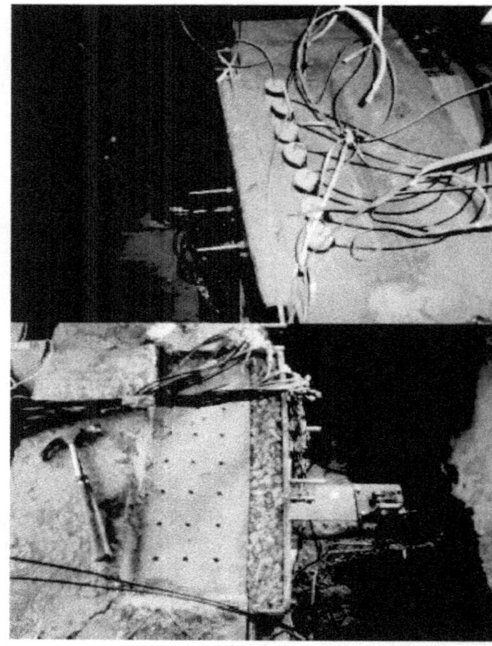

Figure 1. The set-up for the confined test.

There were two set-ups with different confinement conditions.

The free blasts were made without any material at the face of the block.

The confined conditions were created by placing a U-shaped steel plate at the face of the block (Fig. 1). The steel plate was fastened to the yoke by using threaded rods running along the yoke (four on each side), creating a space in front of the burden of a width of around 65 mm. The space was filled with non-magnetic crushed material (debris) to emulate sub-level caving conditions. The debris properties are well defined from the previous studies (Johansson et al. 2008, Johansson & Ouchterlony 2011). It was composed of fragmented cement mortar that followed a Swebrec sieving curve (Ouchterlony 2005). This material emulates the full-scale caved material at the front of a production ring burden i.e. for small scale x_{50} = 8.0 mm, x_{max} = 16.0 mm with a Swebrec undulation exponent of b = 2.0.

The block was charged with a PETN-cord with the strength of 20 g/m, giving a specific charge 2.6 kg/m^3 for the 5 holes design and 4.1 kg/m^3 for the 7 holes design. The coupling ratio was 2.4 based on 1400 kg/m^3 explosive's density. This roll of cord had a Velocity of Detonation (VoD) of around 7540 m/s. The ratio VoD/C_p = 1.98, which is greater than 1.375 for which Blair (2010) states that the Mach waves contain a significant amount of seismic energy that might significantly influence the fragmentation due to preconditioning of the material.

The blastholes were connected to each other by a trunk line and short T-connection lines with the main line i.e. PETN-cord with strength of 5 g/m and VoD about 7290 m/s. The trunk line defined the delay time of each blasthole. The connection between the main charge and the connection line was made by using a modified plastic cylinder with height of 60 mm and 50 mm diameter in which the main charge was embedded 30 mm and the connection line was twisted 3 times to increase the amount of explosive, which is in contact with the main charge. Additionally, this small cylinder had a third hole for a coaxial cable for measuring the actual delay time.

2.2 *Burden compaction measurements*

During the fragmentation tests, dynamic tests were performed to measure the burden compaction and velocity. The instrumentation used embedded 'Nails', accelerometers and a draw wire. The Nail structure contained a plunger (steel), a coax pipe (steel), 10 coaxial cables (RG174/U) and a steel plate for proper mounting on the block's surface (Fig. 2). The application of the method is very simple (Fig. 3).

When the burden starts moving, it pushes the plunger against the coaxial cables, in turn the plunger cuts the cables and an oscilloscope or recording device will record the time, by indicating a spike as is shown in Figure 10. With the obtained time and the distance that the plunger runs along the coax pipe, it is easy to calculate the velocity. The resolution of the method is about 1.7 mm.

The draw wire (Firstmark Controls) is a position transducer which converts the mechanical motion into electrical signal. The draw wire is composed

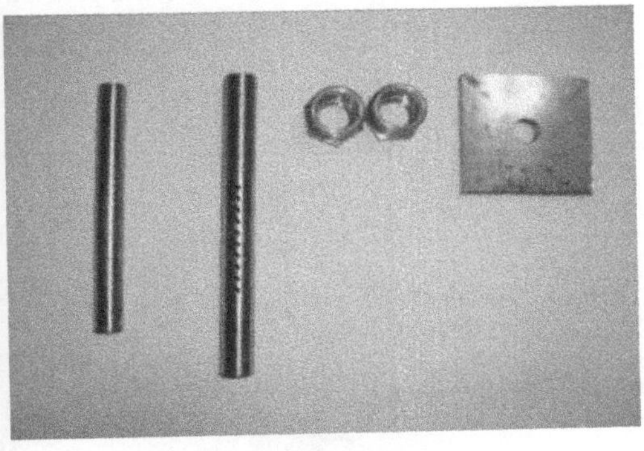

Figure 2. The parts of the Nail method.

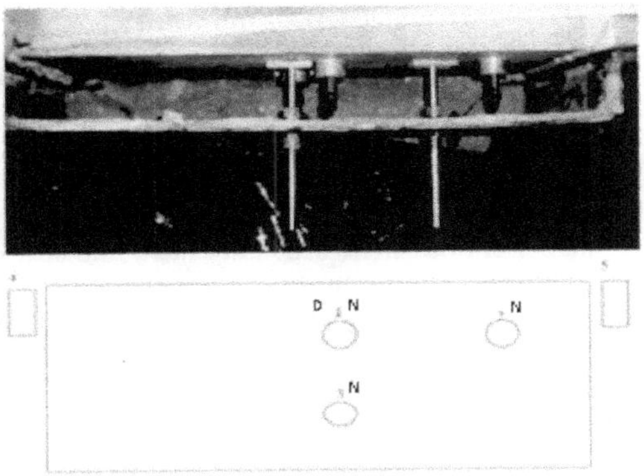

Figure 3. The Nails method and the installation points.

Figure 4. Draw wire.

of a cable, a drum, a spring, a sensor and a signal conditioner. When the burden starts moving, the attached cable on it is going backwards rotating the drum which is tensioned by the internal spring and the sensor produces an electrical output proportional to the displacement cable travel.

The measuring equipment was installed on the face of the magnetic mortar block, in such a way that it formed a triangle. The three installation points on the face were chosen by taking into consideration the symmetry of the face of the magnetic mortar block (Fig. 3). The numbers in Figure 3 show the installation points of the accelerometers, the 'N' denotes the nails position and the 'D' denotes the draw wire position.

The installation point of the draw wire was the same as the accelerometer #1 and the Nail which was located at middle upper point of the face. This was done to acquire comparable results from the three different methods. The draw wire was screwed to a corner (Fig. 4). The corner was modified in order to fit on the steel plate. The draw wire was not installed in the recommended way but in the inverse measuring direction. Instead of pulling out the wire, it was snapped back.

3 THE TEST PROCEDURE

The tests were conducted at a FOI (Swedish Defense Research agency) at the Swebrec's testing facilities in Grindsjön and at the Kimit testing site at the Kiruna mine. The material losses were around 3% because the tests took place in blasting chambers in both sites, which are lined and sealed with blast mats to minimize the losses.

The test procedure of row #1 was to place the U-shaped steel plate in front of the burden creating a space of around 65 mm. The next step was to pack debris in front of the burden of the block to emulate confined conditions. The same procedure was followed for row #2 as well but instead of placing a U-shaped steel plate, a flat plate was installed in front of the new burden. The new burden was not as smooth as in row #1 due to back break, hence the average distance between the burden and the steel was around 70 mm.

Before the preparation of the next round, the loose material was collected and placed in buckets. The broken magnetic mortar material weighed around 32 kg but it contained both debris and part of the yoke. The final weight of the loose material was 55 kg for each test.

The instrumentation consisted of two 8-channel Datatraps (MREL) as the main recorder with sampling rate 10 MHz and a 4-channel Picoscope 4424 with sampling rate 20 MHz, 0.05 μs sample interval and 12-bit resolution. Seven piezoelectric accelerometers were installed. Three accelerometers model DY 3200 BM, range 70 000 g were installed on the surface of the block. In some tests 2 more accelerometers (model B&K 8309, range 100 000 g) were installed on the surface and 2 (model Endevco 7255A-1 Pyrotron, range 50 000 g) were placed in the yoke to both sides of the block in the extensions planes of the rows. The draw wire was manufactured by Firstmark Company with a sensitivity of 139.38 mm/V.

The choice of the delay time was based on previous work (Johansson & Ouchterlony 2012). Blocks #2, #5, #6 & #7 belong to the first design with 5 holes in each row and the rest of them to the second design with 7 holes in each row. The first design was used as reference. The delay time was scaled by using the ratio of the spacing distances between the two designs (110/82.5 = 1.33). Thus, the delay time of the second design was 110 μs and 218 μs corresponding to 146 μs and 290 μs of the first design (Table 1). Two firing patterns were tested i.e. V-shape and sequential.

4 RESULTS AND ANALYSIS

4.1 *Signal analysis*

A typical output of the accelerometers and the initiation system is presented in Figure 5. The P-wave acceleration peaks roughly at 105.000 g. The time windows were not the same in all recorders, this is why the recording time should have been normalized and synchronized with respect to the detonation time of the first blasthole. The detonation time was set to be zero in the analysis.

Figure 5 illustrates signals from Block #3 row #1 (110 μs delay). The black spikes correspond to the initiation, they were used to verify the planned delay time. The recorded signals are the output of the accelerometers which were installed in different known distances from the source point as is shown in Figure 3. Based on this the P-wave velocity can be calculated (Fig. 6). In Figure 6, the P-wave velocities from the tests and for both rows are presented.

It is obvious that the scatter in the above figure is quite high especially for the second row, moreover, the second row has mostly lower values for the P-wave velocity than the first row. This is likely because the second row is preconditioned from the first row, in which the material is intact.

4.2 *Fragmentation*

4.2.1 *Fragmentation under confinement*

There are many ways to evaluate the fragmentation of a blast using different models or techniques. The basic indicators can be the median fragment size

Table 1. The tests plan.

Block	Delay time, μs	Firing pattern	Conditions	No. of holes	Specific charge, kg/m^3
1	110	V-shape	Confined	7	4.1
2	290	Sequential	Confined	5	2.6
3	110	Sequential	Confined	7	4.1
4	218	Sequential	Confined	7	4.1
5	290	Sequential	Free	5	2.6
6	0-R1 73-R2	Sequential	Free	5	2.6
7	290	Sequential	Free	5	2.6

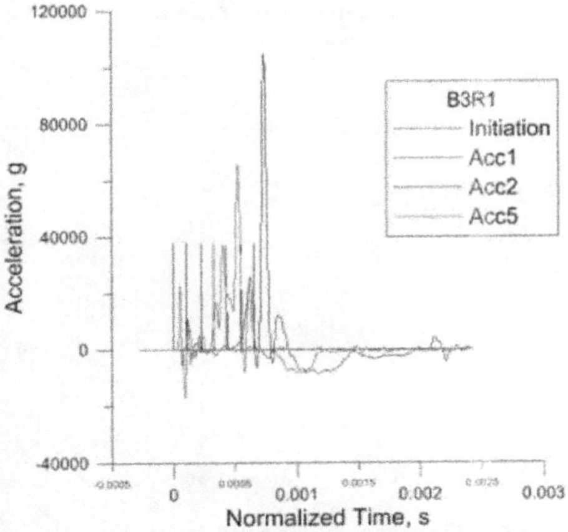

Figure 5. Output signals (Block #3 row #1).

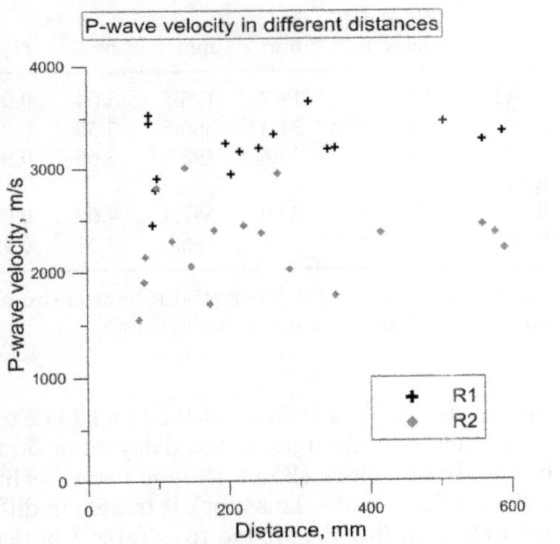

Figure 6. P-wave velocities in different distances from the source point.

(x_{50}) of the material and the maximum fragment size (x_{max}). The value of the median fragment size can be obtained from a sieving process where the total amount of blasted material is sieved. The dry sieving process has some restrictions such as lowest limit of fragmentation e.g. 0.063 mm (depending on the equipment).

The fragmentation model used was the 3-term Swebrec function (Ouchterlony 2005):

$$P(x)=\frac{1}{1+\left[\dfrac{\ln(x_{max}/x)}{\ln(x_{max}/x_{50})}\right]^{b}} \quad (1)$$

where x_{max} = maximum fragment size; x_{50} = median fragment size; and b = undulation parameter.

After the sieving analysis of the loose material, both magnetic mortar and debris, the data were processed using the above function. Table 2 presents the x_{50}, x_{max} and the undulation exponent b, of the Swebrec function of the blocks under confined conditions. The fit of the function is very good i.e. the determination coefficient is greater than 0.9987. The fragmentation characteristics were calculated based on Equation 1.

The data of Table 2 are illustrated in Figure 7. For instance, B1R1 denotes block #1, row #1 etc. A direct comparison is not possible to make since

Table 2. Fragmentation characteristics under confinement.

	Nominal delay, μs	x_{50}, mm	x_{max}, mm	b	r^2
B1R1	110	33.0	67.9	1.79	0.9988
B2R1	290	35.9	529.4	4.98	0.9994
B3R1	110	29.3	123.4	2.47	0.9992
B4R1	218	19.7	65.7	2.14	0.9994
B1R2	110	21.5	86.5	2.60	0.9987
B2R2	290	23.5	211.2	3.48	0.9997
B3R2	110	15.7	94.5	2.75	0.9988
B4R2	218	15.8	88.0	2.70	0.9993
B1R3	110	13.3	82.6	3.00	0.9996
B3R3	110	11.0	66.5	2.68	0.9989

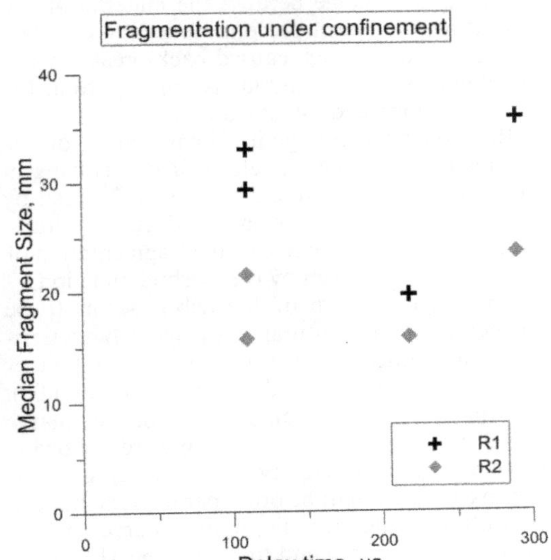

Figure 7. The fragmentation results under confinement (delay time at 290 μs belongs to the first design, 2.6 kg/m³).

several parameters are changed in the different designs such as firing pattern, specific charge and delay time. Thus, the comparison of the results must be carefully made between particular blocks.

Comparisons can be made between the first rows of the following blocks:

- Block #1 and Block #3. The sequential pattern gives a 11% finer fragmentation than the V-shaped pattern,
- Block #2 and Block #4 due to different specific charge, which results in a finer fragmentation by around 45% for the overcharged block,
- Block #3 and Block #4 due to different delay times, the block with the longer delay time (218 μs) gives a finer fragmentation, by almost 33%, than the block with 110 μs delay time,
- Block #2 and Block #5 due to different confinement conditions, the free face block gives a finer fragmentation, by 45%, than the block under confined conditions.

Regarding the second row similar comparisons can be made: the sequential initiation gives almost 27% finer fragmentation than the V-shaped pattern, the overcharged block gives 32%, finer fragmentation and finally, the blocks with different delay times give almost the same median fragment size. Possible explanations are the smaller burden due to uneven formed free face or a different crack pattern caused by the blast of the first row.

Generally speaking, the median fragment size of the first row is coarser than the median fragment size of the second row.

This is reasonable because the material of the second row has been preconditioned by the blast of the first row, which caused back breakage and radial cracks in the material as opposed to at the first row, which had intact material.

Regarding the change in fragmentation of the confined material i.e. the effect of debris lying in front of the burden, it seems to be influenced by the changed parameters in the different setups. Figure 8 shows the change of the fragmentation of the debris as it is given by the Swebrec function.

The fragmentation of the debris seems to be subjected to the Natural Breakage Characteristic (NBC) concept as it was proposed by Steiner (1991, 1998) and widened by Moser et al. (2000). This means that by changing the specific charge the distribution curves move upwards showing more fine material. However, in these tests, as was observed there might be other parameters that can possibly influence the distribution curve such as the initiation pattern and the delay time (Fig. 8).

4.2.2 *Fragmentation under free face*

A series of tests were performed using the same magnetic mortar block setup. However, in this case there is no material in front of the block. The only parameter that changes is the delay time of the blocks. In one block (Block #6) the two rows have different delay times. This block is treated in different way regarding the second row. Table 3 presents the fragmentation characteristics given by the Swebrec function for both rows.

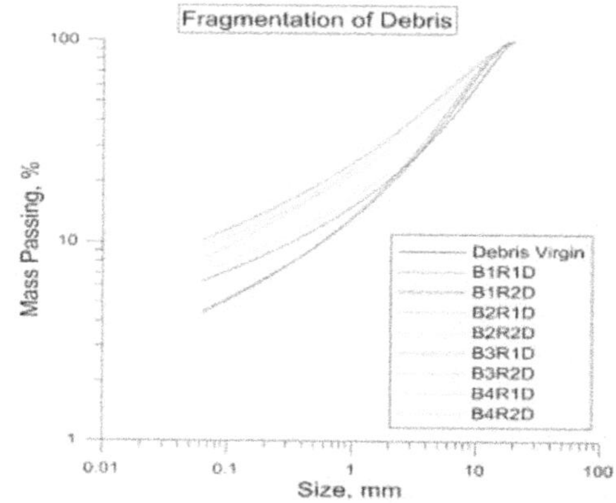

Figure 8. The fragmentation results of the debris.

Table 3. Fragmentation characteristics under free face.

	Nominal delay, μs	x_{50}, mm	x_{max}, mm	b	r^2
B5R1	290	19.7	126.5	3.04	0.9993
B6R1	0	31.3	366.6	3.55	0.9997
B7R1	290	22.6	192.8	3.66	0.9996
B5R1*	–	–	–	–	–
B6R2	73	25.9	587.4	4.69	0.9995
B7R2	290	17.5	156.9	3.23	0.9996

* There are no results for Block #5 due to extensive back breakage from the blasting of the first row.

Figure 9 illustrates the median values of the fragment size of the blasted material. As is observed in Figure 7, the second row gives smaller median size than the first row. The instantaneous initiation gives coarser fragmentation than the rows with delay time as was also observed by Johansson & Ouchterlony (2012).

4.3 *Burden compaction*

The second part in the study was to evaluate the burden behavior during the blast. As reference method was chosen to be the Nails method. This was because the starting point of the burden movement was hidden behind the reflections of the waves, recorded with the accelerometers, in the block. This makes it almost impossible to separate

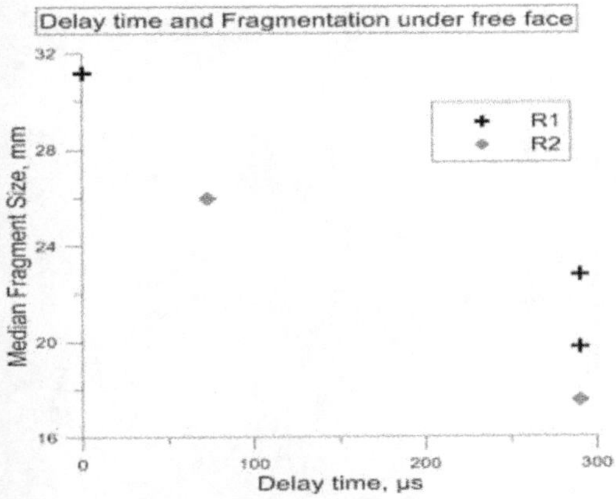

Figure 9. The fragmentation results under free face.

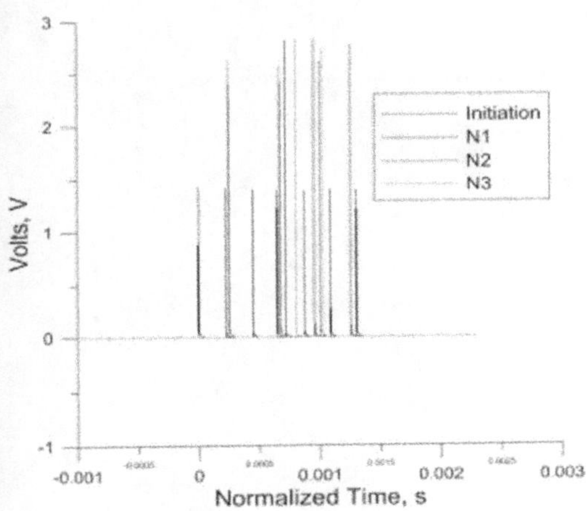

Figure 10. Signals from nails (Block #4 row #1).

exactly the part of the waveform that corresponds to the burden movement. The alternative methods implemented was the use of draw wire, however the draw wire should be installed in the reverse way to how it usually works i.e. instead of pulling out the wire, it should snap it back, also the draw wire showed a very slow response, which made it unsuitable for this kind of small scale tests.

The compaction is given as a range because the method does not provide a clear-cut value. The resolution of the method is 1.7 mm (distance between two consecutive holes) and the plunger can stop somewhere between the holes.

Hence, the exact position of the plunger is not clear. Figure 10 shows the recorded spikes of the initiation system and the nails from different positions. The procedure that was followed to calculate the compaction and the velocity of the burden is to find the point where the burden starts moving. This is the first spike of the Nails and to measure the time span between two consecutive spikes of the Nails. After that, by knowing the time and the distance (compaction) of the two spikes, the velocity can be calculated. In Figure 10 the burden starts moving after 250 µs after the detonation of the first blasthole.

Table 4 presents the results of the nails in different positions on the face of a block. It is obvious that the scatter corresponding to compaction is quite large. One reason is that the plunger could get stuck in the coax pipe due to dust. Additionally, some data are missing due to problems with the recording device and in two cases the proper installation of the equipment was not possible due to very rough face of the second row.

The processed data are presented in Table 5, where the maximum compaction peaked at 12.6 mm and the peak velocity at 29 m/s. The behavior of the burden during the blast is portrayed in Figure 11. At the beginning the velocity gradually increases until reaching the peak value, after that the velocity abruptly drops down and finally begins to level out approaching the peak compaction. With respect to the results, there is evidence that shows that the burden movement stops at a lower value than the peak compaction, which means that the burden moves few millimeters backwards, the exact distance is not available because the nails were not able to record this behavior (Fig. 14).

Table 4. The results of the nails.

	Compaction, mm		
	1	2	3
B1R1	12.6–14.3	12.6–14.3	–
B1R2	–	10.9–12.6	12.6–14.3
B2R1	6.7–8.4	10.9–12.6	10.9–12.6
B3R1	10.9–12.6	6.7–8.4	12.6–14.3
B3R2	2.5–4.2	10.9–12.6	6.7–8.4
B4R1	12.6–14.3	10.9–12.6	6.7–8.4

Table 5. The burden movement.

Compaction, mm	Velocity, m/s	Standard deviation of the velocity
0.0	0.0	0.0
4.2	17.9	6.6
6.7	28.8	4.5
8.4	18.6	6.2
12.6	8.62	4.3

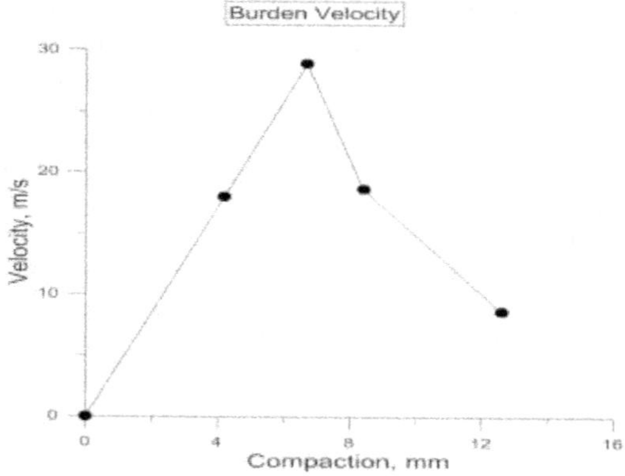

Figure 11. Burden velocity.

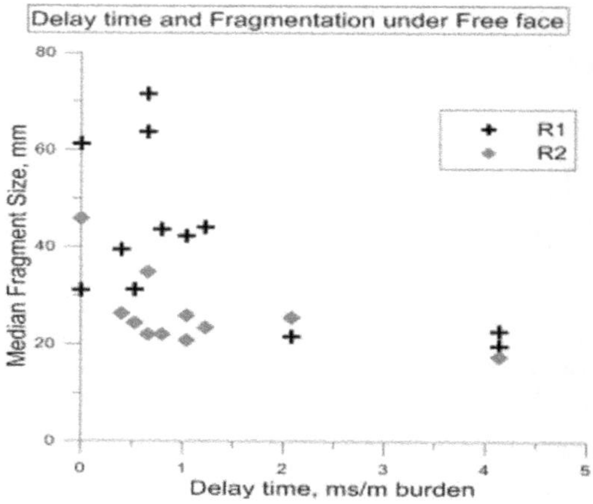

Figure 12. Median fragment size under free face (including data from Johansson & Ouchterlony 2012).

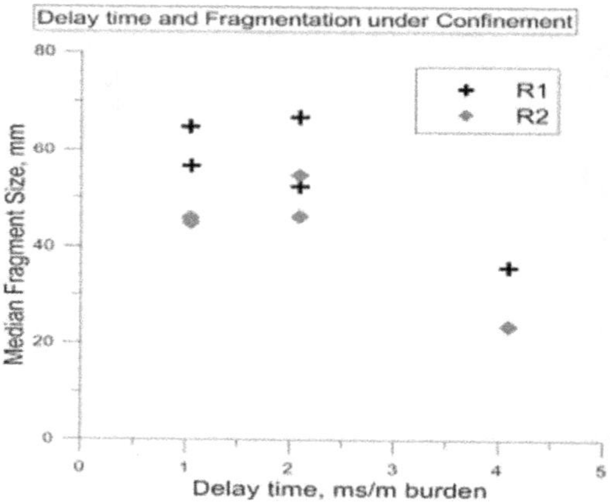

Figure 13. Median fragment size under confinement (including data from Johansson & Ouchterlony 2012).

5 DISCUSSION

5.1 *Fragmentation*

Combining data from previous tests (Johansson & Ouchterlony 2012) with the data from these tests result in Figures 12 and 13. There is a tendency of the median fragment size to become smaller when using delay times up to 4.1 ms/m of burden. Previous researchers (Onederra 2008, Stagg & Rholl 1987, Stagg & Nutting 1987) have presented that the fragmentation changes when the delay time increases. They have especially noticed that the optimum fragmentation depends on the delay time and the fragmentation reaches a minimum or makes a plateau at particular span of delay time. Figures 12 and 13 illustrate the median fragment size under free and confined conditions. In both of the figures, there is a trend that shows that the median fragment size becomes finer for longer delay times.

For the free face case the delay time ranges from 0 to 4.1 ms/m of burden and for the confined conditions ranges from 1 to 4.1 ms/m of burden equivalent to 73 μs and 290 μs respectively. For the longest delay time 4.1 ms/m (290 μs) under free face conditions, the median fragment size for the row #2 is missing because of the extensive back break damage of the block from the blast of the row #1. The extensive back break damage was observed in the blocks with delay times longer than 3.74 ms/m (218 μs) and it might have an influence on the fragmentation.

In the cases of the longer delay times 3.74 & 4.1 ms/m of burden (218 μs & 290 μs), the radial crack reaches roughly 120 & 160 mm from the first blasthole or the previous detonated blasthole, distances which are longer than the spacing between the blastholes, with a crack propagation velocity to be between 500–600 m/s (Johansson & Ouchterlony 2012) which means that the crack will have reached the neighboring blasthole before the detonation of it.

Consequently, the stresses induced by the shock wave and the gases of the second blasthole may enhance the number and length of crack growth.

To test longer delays than 4.1 ms/m of burden (290 μs) is not possible with this test setup because each hole acts independently from the others, which means it is out of any kind of potential interactions of different mechanisms when the delay time is longer. Additionally, the burden starts moving after roughly 250 μs after the initiation of the first blasthole. Thus, there is a natural restriction of testing interactions for longer delay times with this setup.

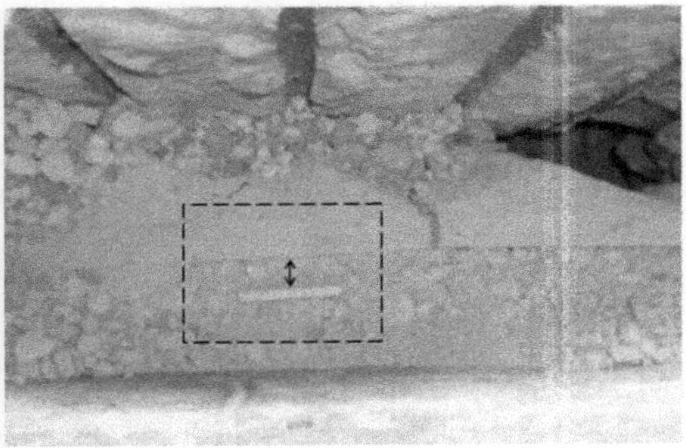

Figure 14. The burden after blast.

5.2 *Burden*

The literature is very limited regarding the burden behavior in small scale under confined conditions.

According to the tests the burden reaches a velocity of around 29 m/s. The V-shaped initiation pattern gives about 15% higher compaction than the sequential initiation pattern.

The maximum compaction reaches about 20% of the total distance between the face of the block and the steel plate. The burden does not stop at the peak point (maximum compaction) but it moves backward few millimeters, at least in small scale. A possible explanation for this is that the compacted material expands due to relaxation as the loading force by blast dissipates.

A sign of that is shown by the small steel plate, in Figure 14, which has stopped at the maximum point of the compaction, however the nail method was not design to record backward movement of the burden. Alternative methods such as optical or magnetic methods that are able to record this behavior could be used.

After the burden stops, it is not in contact with the newly formed face of the block. There is a small air gap between the blasted burden and the new face, which is filled with the debris that was placed in front of the block's face. It is obvious that a small air gap was formed and maintained its shape behind the blasted burden before it was filled by the debris.

Finally, there was not any significant difference of compaction between the first and the second row, 11.2 ± 2.3 mm and 11.1 ± 2.0 mm respectively.

6 CONCLUSIONS

These small scale tests showed the behavior of the fragmentation and the burden under different confinement conditions.

The P-wave velocity is higher in the intact magnetic mortar (row #1) than the preconditioned material (row #2).

The fragmentation under confined conditions is coarser than under free face conditions as was also observed by Johansson & Ouchterlony (2011).

The fragmentation becomes finer for longer delay times than for very short delay times.

The fragmentation of the debris changes when changing fundamental blasting parameters such as the specific charge, the delay time and the initiation firing pattern.

The Swebrec function was used to analyze the fragmentation results of the magnetic mortar and the debris giving very good fit.

The nail method gave reasonable results for the burden behavior as opposed to the draw wire and the accelerometers results where it was almost impossible to extract the burden movement.

The burden peaks at a velocity of around 29 m/s and then comes few millimeters backwards, before it stops in its final position, at least in small scale tests.

The compaction is roughly the same for the first and the second row.

The compaction depends on the firing pattern. The V-shaped firing pattern gives higher compaction by 15% due to differential movement of the two parts of the burden.

Finally, there are some mechanisms active during long delays (4.1 ms/m of burden or 218 μs) causing extensive back break damage that can ruin the block, in small scale tests. They show a dependency of the delay times and the back break. Further study of these mechanisms is required to identify the possible interactions (crack propagation and gas expansion).

ACKNOWLEDGEMENT

This work has been financed by LKAB. The authors wish to acknowledge the Swebrec's researcher Ulf Nyberg, the LKAB's manager in mining engineering and technology Anders Nordqvist, the sieving laboratories in Kiruna and in Boden, LKAB's sieving laboratory and Väglaboratoriet i Norr AB respectively. Finally, we would like to thank Kimit for letting us use their testing site.

REFERENCES

Blair, D.P. 2010. Seismic radiation from an explosive column. *Geophysics* 75(1): E55–E65.

Firstmark Controls. Rugged. Miniature Position Transducers. Solution Guide.

Johansson, D. 2008. Fragmentation and waste rock compaction in small scale confined blasting. Lic. thesis 2008:30. Luleå Univ. Techn. Luleå, Sweden, 117 pp+app.

Johansson, D, & Ouchterlony, F. 2011. Fragmentation in small scale confined blasting. *Int. J. Mining and Mineral Engineering* 3(1): 72–94.

Johansson, D. & Ouchterlony, F. 2012. Shock wave interactions in rock blasting-the use of short delays to improve fragmentation in model scale. *Rock Mechanics and Rock Engineering*. In press. DOI 10.1007/s00603-012-0249-7.

Johansson, D., Ouchterlony, F. & Nyberg, U. 2007. Blast against aggregate confinement, fragmentation and swelling in model scale. In P. Moser (ed.), *Proc. 4th EFEE World Conf. on Explosives and Blasting, EFEE, Vienna, Austria*, pp. 13–27.

Johansson, D., Ouchterlony, F., Edin, J., Martinsson, L. & Nyberg, U. 2008. Blasting against confinement, fragmentation and compaction in model scale. In H. Schunnesson & E. Nordlund (eds), *MassMin 2008—Proc. 5th Int. Conf. & Exhib. on Mass Mining*. Rotterdam: Balkema, pp. 681–690.

Katsabanis, P.D., Tawadrous, A., Braun, C. & Kennedy, C. 2006. Timing effects on the fragmentation of small scale blocks of granodiorite. *Fragblast—Int. J. Blast Fragm.* 10(1–2): 83–93.

Moser, P., Cheimanoff, N., Ortiz, R. & Hochholdinger, R. 2000. Breakage characteristics in rock blasting. In R. Holmberg (ed.), *Proc. 1st EFEE Conference on Explosives & Blasting Technique*. Rotterdam: Balkema, pp. 165–170.

Onederra, I. 2008. A delay timing factor for empirical fragmentation models. *IMM trans. Mining Technology* 116(4): 176–179.

Ouchterlony, F. 2005. The Swebrec function: linking fragmentation by blasting and crushing. *Mining Technology (Trans. Inst. Min. Metall. A)* 114: A29–A44.

Rossmanith, H.P., Uenishi, K. & Kouzniak, N. 1997. Blast wave propagation in rock mass. Part I: Monolithic medium. *Fragblast—Int. J. Blast Fragm.* 6(1): 317–360.

Stagg, M.S. & Nutting, M.J. 1987. Influence of blast delay time on rock fragmentation: one tenth scale tests. *BuMines IC 9135*, Chicago, USA, pp. 79–95.

Stagg, M.S. & Rholl, S.A. 1987. Effects of accurate delays on fragmentation for single row blasting in a 6.7 m bench. *Proc. 2nd Intnl. Symp. on Rock Fragmentation by Blasting, Keystone, Colorado*, pp. 210–230.

Steiner, H.J. 1991. The significance of the Rittenger equation in present-day comminution technology. *Proc. XVII Int. Minerals Processing Cong, Dresden*, Vol 1, pp. 177–188.

Steiner, H.J. 1998. Zerkleinerungstechnische Eigenschaften von Gesteinen. Felsbau 16: 320–325.

Vanbrabant, F. & Escobar, A.E. 2006. Impact of short delays sequence on fragmentation by means of electronic detonators: theoretical concepts and field validation. *Proc. 8th Int. Symp. on Rock Fragmentation by Blasting—Fragblast 8, Santiago, Chile*, pp. 326–331.

Size distributions

Measurement and Analysis of Blast Fragmentation – Sanchidrián & Singh (Eds)
© 2013 Taylor & Francis Group, London, ISBN 978-0-415-62140-3

A three parameter rock fragmentation distribution

A.T. Spathis
Orica Mining Services Technical Centre, Kurri Kurri, NSW, Australia

ABSTRACT: The traditional prediction of the rock fragmentation distribution produced by blasting relies on the Kuz-Ram model. A recent development has been the Swebrec function, a three parameter model that appears to fit measured mass percent passing distributions at least as well as the Weibull distribution embodied in the Kuz-Ram model. It has the added benefit of fitting the fines region well. A generative or mechanistic model that produces the cumulative mass percent passing curve for the Swebrec function remains to be found. An alternative three parameter model is proposed that combines the Weibull distribution for the larger sizes and mimics some of the behaviour of the Swebrec function for the finer sizes using an exponential distribution. The fines are important in "heavy blasting" for comminution. The new model is compared to various other potential models using several data sets and is shown to be only slightly poorer to the fits produced by the Swebrec function. The three parameter rock fragmentation distribution model proposed has a simple algebraic analytical expression for the mean, unlike the Swebrec function, and as such can be readily embedded into a Kuz-Ram-like approach for predicting the rock fragmentation distribution produced by blasting. Furthermore, the fragmentation distribution produced by the model can be explained by two physical mechanisms related to the exponential and Weibull distributions.

1 INTRODUCTION

The Kuz-Ram model estimates the fragmentation distribution of the broken rocks produced by blasting. It is widely used by blasting engineers partly because it is readily coded into a spreadsheet and partly because it provides satisfactory predictions after calibration (Singh et al., 2006). It uses the Kuznetsov (1973) estimate for the mean size and the Rosin-Rammler or Weibull curve for the distribution (Cunningham, 1983, 1987). The Rosin-Rammler curve is a two-parameter, unimodal probability density distribution that uses a characteristic size for the central tendency and a uniformity index for the spread. Cunningham's contribution was to suggest how these parameters could be estimated from blast design parameters.

Spathis (2009) surveyed formulae and techniques for assessing features of blast-induced fragmentation distributions. It is clear from that work that there are a disparate range of methods and approaches, some theoretical and some based on curve-fits to measured data. The Weibull curve is used to fit fragment size distributions, including those from sieved rock pile data and from distributions based on image analysis data (Singh et al., 2006). The model underestimates the fines region and alternative approaches have been proposed (Djordjevic, 1999; Thornton, et al., 2001; Esen et al., 2003).

Ouchterlony (2005) introduced a different curve to model the shape of the fragmentation distribution. He called it the Swebrec function. It fits a wide range of measured fragmentation data and, in particular, fits the fines region data well when the coarser size fractions are fitted. Ouchterlony (2005) proposes an analogous model to the Kuz-Ram model whereby the Rosin-Rammler curve is replaced by the Swebrec function and the Kuznetsov (1973) estimate is used again for the mean size estimate. Unfortunately, Ouchterlony (2005) and Ouchterlony and Moser (2006) confuse the mean and the median values for a distribution. Spathis (2004) identified such a discrepancy in connection with the original Kuz-Ram model, and indicated that others in blasting had also made that mistake. The essential point is that the Kuznetsov (1973) expression is for the mean size and not the median size. These are only identical, in general, when the distribution is symmetric.

The Kuz-Ram model and the Swebrec function are revisited. The Swebrec function has a limitation relating its form and including the singularity at sizes approaching zero. It appears that an algebraic analytical expression for the mean size of the Swebrec function is unavailable. Hence, the simple replacement of the Rosin-Rammler distribution by the Swebrec function in the Kuz-Ram model is not possible without further consideration. Some suggestions are canvassed in this paper. Alternative cumulative distribution functions that have three parameters are also considered.

Recent work by Sanchidrian et al. (2009) considers a number of distribution functions and their so-called two-component versions whereby two identical single distribution functions are combined to produce a better fit to measured fragmentation data. In contrast, the present work focuses on distribution functions that have just three parameters, with some of the functions combining two distributions of different character. A further emphasis in this paper is the desire to have a distribution function that may arise from physical principles or mechanistic considerations.

1.1 *Rock fragmentation distribution functions*

The definitions used for the continuous distributions in this paper are taken from Evans et al. (2000) and are shown in Table 1.

A series of Cumulative Distribution Functions (CDF) are presented in this section. Firstly, a description is given of the Kuz-Ram model (Cunningham, 1983, 1987). It is currently the most common fragmentation model used in blasting and is based on the two-parameter Rosin-Rammler or Weibull function. The three-parameter Swebrec model (Ouchterlony, 2005) is discussed next as it appears to fit a wider range of sizes than does the Kuz-Ram model. The sigmoidal fit used by Blair (2004) is described. Finally, a series of alternative three-parameter CDF models are discussed: a three-parameter modified Rosin-Rammler or Weibull function, a combined exponential/Weibull function and, lastly, a combined Rayleigh/Weibull function.

1.2 *Kuz-Ram model*

The Kuz-Ram model for the fragmentation distribution of rocks produced by blasting melds the Kuznetov (1973) expression for the mean size and the Rosin-Rammler curve for the shape of the distribution. The approach found prominence in blasting when Cunningham (1983, 1987) provided a formula for the dispersion of the distribution in terms of blast design geometry.

The CDF of a Rosin-Rammler or Weibull distribution (Evans et al., 2000) describes the cumulative percent passing, $F(x)$, as,

$$F(x) = 100\left(1 - \exp\left[-\left(\frac{x}{x_c}\right)^n\right]\right) \tag{1}$$

where the two parameters that describe the distribution are the characteristic size, x_c, and the uniformity index, n. An example of the normalised distribution is shown in Figure 1.

The mean size of a Weibull distribution described above is given by,

$$\mu = x_c\Gamma\left(1 + \frac{1}{n}\right) \tag{2}$$

where Γ is the gamma function, and the median or 50% passing size is given by,

$$x_{50} = x_c\left(Log_e 2\right)^{1/n} \tag{3}$$

Kuznetsov (1973) discussed the application of the Rosin-Rammler distribution to the fragmentation

Table 1. Definition of distributions.

Term	Symbol	Description
Distribution function (DF) or cumulative distribution function (CDF)	$F(x)$	$F(x)$ is the probability that the size takes a value less than or equal to x. $F(x) = \Pr[X \le x] = \alpha$ $F(x) = \int_{-\infty}^{x} f(u)\,du$ where $f(u)$ is defined below.
Probability density function (PDF)	$f(x)$	A function whose general integral over the range x_L to x_U is equal to the probability that the variate takes a value in that range. $\int_{x_L}^{x_U} f(x)\,dx = \Pr[x_L < X \le x_U]$ $f(x) = \frac{d(F(x))}{dx}$
Median	x_{50}	Value of x such that $F(x) = 0.5$.
Mean	μ	$\mu = \int_{-\infty}^{+\infty} x\,f(x)\,dx$

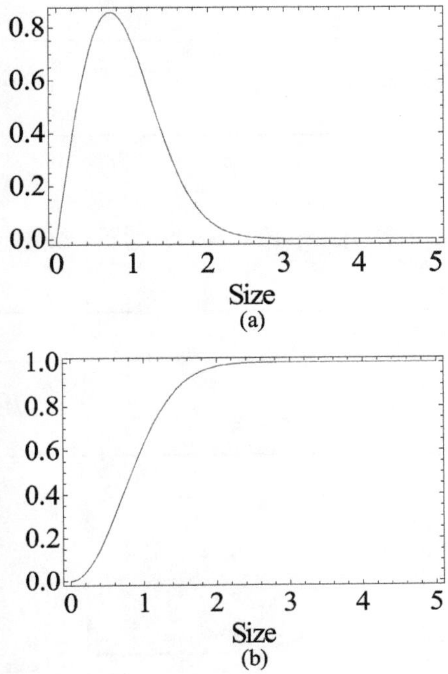

Figure 1. Weibull Fragmentation size distribution shown as a normalised (a) PDF and (b) CDF. The distribution in this case is a normalised distribution with a characteristic size = 1 and a uniformity index = 2. Scaling the CDF produces the percent passing curve of Equation (1).

sizes of blasted rock. Analysing both laboratory scale and field scale blast results, he found an empirical formula that fit the average fragment size, d, in centimetres:

$$d = A(V_0/Q)^{4/5} Q^{1/6} \tag{4}$$

Where A is the rock factor and equal to 7 for medium rock, equal to 10 for hard, fissured rock and equal to 13 for hard, competent rock. V_0 is the volume of rock broken per hole. Q is the mass of TNT of equivalent energy to the explosive in the borehole (kg) and is equal to 0.87 kg per kg of ANFO.

Cunningham (1983, 1987) developed the Kuz-Ram model for predicting the cumulative size distribution of fragments produced by blasting. The model uses the Rosin-Rammler (Weibull) equation for the shape of the distribution where the mean size (d) is given by the Kuznetsov equation above. The other parameter that defines the Rosin-Rammler curve is the uniformity index and in original form is given by Cunningham (1983) as,

$$n = (2.2 - 14\, diam/B)(1 - W/B)((1+R)/2)L/H \tag{5}$$

where B is the burden (m), $diam$ is the hole diameter (mm), W is the standard deviation of drilling accuracy (m), R is the spacing to burden ratio, L is the charge length above grade level (m) and H is the bench height (m). If a staggered pattern is employed, n is increased by 10%.

The modified version for the uniformity index (ignoring multiple charges) is given by (Cunningham, 1987),

$$n_{\text{mod}} = (2.2 - 14\, diam/B)(1 - W/B)\sqrt{(1+R)/2\ L/H} \tag{6}$$

and this introduces a square root function to diminish the effect of the spacing to burden ratio on the uniformity index.

Cunningham (1987) provided a modified rock factor used in the Kuznetsov (1973) expression for the mean size (Equation (4)). It is based on the "blastability" concept of Lilly (1986) and is given by,

$$A = 0.06(RMD + JF + RDI + HF) \tag{7}$$

The parameters of Equation (7) in brackets are defined in Table 2.

An estimate for the fragmentation distribution using the Kuz-Ram model may be made as follows:

1. Estimate the mean size using the Kuznetsov expression in Equation (4) using either Kuznetsov's original values for the rock factor or those suggested by Cunningham as given in Equation (7).
2. Evaluate the uniformity index using either the original approach in Equation (5) or the modified approach in Equation (6).
3. Evaluate the characteristic size using Equation (2).
4. Use Equation (1) to estimate the percent passing curve based on the above calculations for the characteristic size and uniformity index.

An interesting feature of the Kuz-Ram model is that the mean size is a function of the explosives and rock properties used while the uniformity index is a function of the blast design geometry. Ouchterlony (2009) suggests through a study of rock fragmentation using dimensional analysis developed from asteroid collision problems, that the nature of such a split of effects is inappropriate and perhaps incorrect. Noy (2012) suggested that it is important to understand the sensitivity of the Kuz-Ram model to the various blast design parameters, particularly in Equations (2), (5) and (6). Some of aspects of this sensitivity are discussed briefly here.

For a given uniformity index, the characteristic size is proportional to the mean size provided by the Kuznetsov formula (Equation (4)). In turn, the mean size is directly proportional to the rock factor,

Table 2. Rules for blastability parameters.

RMD	Rock mass description	Powdery/friable vertically jointed massive	10 JF 50
JF	Joint factor = JPS + JPA JF vertical joint spacing where MS is maximum and DP is drilling pattern size (DP > MS)	0.1 m 0.1 m to MS MS to DP	10 20 50
	JPA joint plane angle	dip out of face strike perp. to face dip into fac e	20 30 40
RDI	Rock density influence	RDI = $25r_{rock} - 50$	
HF	Hardness factor E = Young's modulus in GPa. UCS = ultimate compressive strength in MPa	if E < 50 GPa if E > 50 GPa	E/3 UCS/5

and is almost inversely proportional to the powder factor (kilograms of explosive per cubic metre of rock broken per hole). Since the variability of powder factors at a given site vary relatively slowly (say 20%), it is likely that the varying geology in different geotechnical domains within a given pit will be the most significant influence on the Kuznetsov mean size.

Another source of sensitivity on the characteristic size arises from the original work of Cunningham (1983) where he equates the 50% passing size (or median) for a Rosin-Rammler distribution to the Kuznetsov mean fragment size. That is, he mistakenly assumes that the mean and the median are equal for a Rosin-Rammler distribution. Equations (2) and (3) show that this can only occur when the uniformity is infinite and all particles have the same size. For practical purposes, the Cunningham error in the characteristic size is less than about 5% for n > 2, but becomes more significant for n < 2 (Figure 2). The Rosin-Rammler curve becomes more symmetric for values of the uniformity index greater than two. Spathis (2004) identified this error in the original Kuz-Ram model and suggested that it may explain, in part, the inability of the Kuz-Ram model to predict finer size fractions observed in actual blasting. It is a simple matter to correct the original Kuz-Ram model and obtain the characteristic size based on the Kuznetsov estimate of the mean by combining Equations (2) to (4).

If we assume perfect drilling accuracy and a staggered pattern where the spacing to burden ratio is 1.15, then the uniformity index becomes a function of the ratios of diameter to burden, and of charge length to bench height. These two dependencies are illustrated in Figure (3).

1.3 *Swebrec model*

Ouchterlony (2005) and Ouchterlony and Moser (2006) introduced an alternative cumulative size distribution function that fits fragmentation data from blasts and other sources such as crushing and grinding. They called it the Swebrec function. The scaled cumulative percent passing curve, $F(x)$, is,

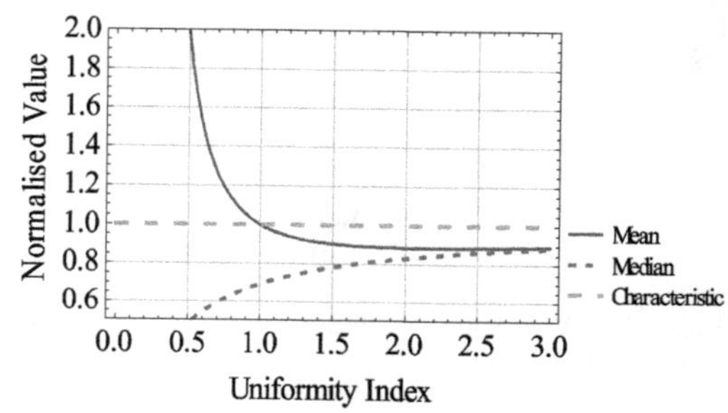

Figure 2. Ratio of the mean and median size to the characteristic size of the Kuz-Ram model versus the uniformity index (Equations 2 and 3).

$$F(x) = 100\left(1/\left\{1+\left[Log_e\left(x_{\max}/x\right)/Log_e\left(x_{\max}/x_{50}\right)\right]^b\right\}\right) \quad (8)$$

where $0 < x \leq x_{max}$ x_{50} is the 50% passing size or median, x_{max} is the maximum fragment size, and b is an undulation parameter. It should be noted that the Swebrec probability density function has a singularity at $x = 0$. A consequence of that singularity and the form of the Swebrec function is that the mean size is not expressed readily as an algebraic analytical function. An example of the Swebrec PDF and the associated cumulative distribution function is shown in Figure 4. The probability density distribution shows the effect of the singularity for small sizes less than say 0.05 of x_{max}. The distributions alter character as a function of the parameter, b (Figure 5), particularly for smaller sizes. As b increases, the probability density distribution becomes more symmetric about the median size if we ignore the singularity that exists as we approach smaller sizes.

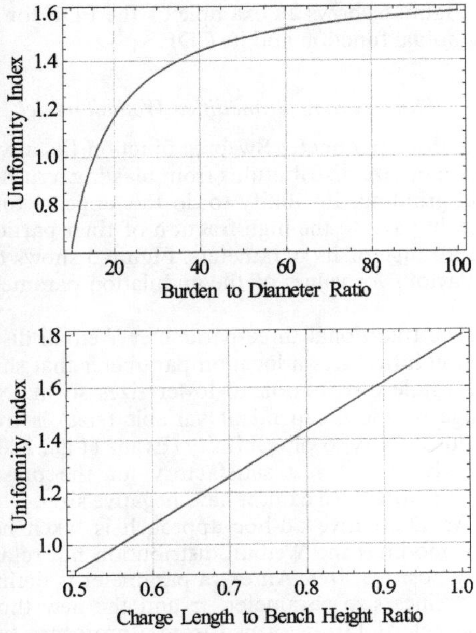

Figure 3. Variation of the uniformity index with (a) burden to diameter ratio and (b) charge length to bench height using Equation (6) with spacing to burden ratio of 1.15 (equilateral staggered pattern) and zero error in drilling accuracy.

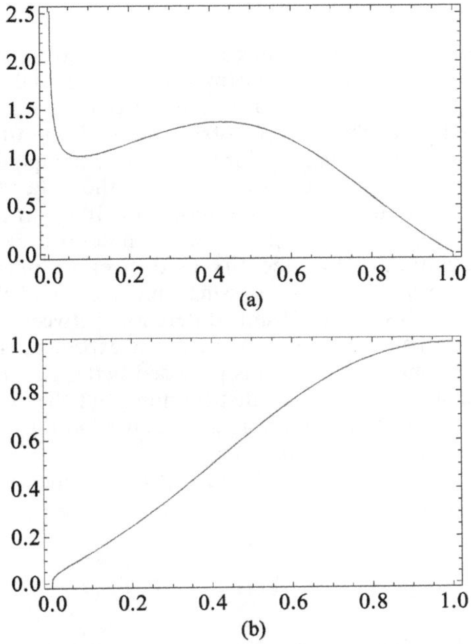

Figure 4. Swebrec function (a) PDF shown as a normalised frequency distribution and (b) its CDF. The distribution in this case is for a normalised Swebrec function with $x_{max} = 1$, $x_{50} = 0.4$, and $b = 2$. Scaling the cumulative distribution produces the percent passing curve of Equation (8).

Ouchterlony (2005) proposed that the Swebrec function replace the Rosin-Rammler distribution in the Kuz-Ram model. He noted that in his data the slopes of a Rosin-Rammler curve and that of the Swebrec function were similar near x_{50}. Equating the two slopes at x_{50} gives,

$$b = 2n\,Log_e(2)Log_e\left(\frac{x_m}{x_{50}}\right) \quad (9)$$

Ouchterlony (2005) suggests that x_m can be chosen as the minimum of the in situ block size, the spacing or the burden of the blast design. Finally, he equates the median size to the Kuznetsov mean size—a strictly incorrect approach as has been mentioned above.

The process for using the Swebrec function to estimate the fragmentation distribution as espoused by Ouchterlony (2005) is as follows:

1. Estimate the mean size using the Kuznetsov expression in Equation (4) using either Kuznetsov's original values for the rock factor or those suggested by Cunningham as given in Equation (7). Use this value for the median size, x_{50}.
2. Evaluate the uniformity index, n, using either the original approach in Equation (5) or the modified approach in Equation (6).
3. Evaluate the maximum size, x_m, from the minimum of the largest in situ block size, spacing or burden of the blast design.
4. Evaluate the undulation parameter, b, from Equation (9).
5. Use Equation (8) to estimate the percent-passing curve based on the above calculations for the maximum size, median size and the undulation parameter.

The key fault with this procedure is that the Kuznetsov mean size is used as the median size in the Swebrec function. Unfortunately, we cannot equate the Kuznetsov estimate of the mean size to the mean size of the Swebrec function since an algebraic analytical expression does not appear to exist for it. Another consideration is that the procedure offered by Ouchterlony requires that the slopes of the Swebrec function and the Rosin-Rammler CDF at the median size are equal, something that should be tested. Indeed, examining the CDF and the PDF of the Swebrec model in Figure 5 it may be seen that at the median size of 0.4 in the CDF of the Swebrec function, the slope increases faster as the undulation parameter decreases from five to unity. The PDF value at the median size shows this effect as it should since the PDF is the derivative of the CDF.

1.4 *Sigmoidal model*

Blair (2004) used three different distributions Rosin-Rammler, log-normal and sigmoidal) to fit fragmentation data obtained from single-hole blasting in three blocks of granite rock He fitted both single distributions and two-component distributions of the same type to the data. The data spanned sizes from about 0.1–100 mm and 0.01–10 kg, or approximately three orders of magnitude. Also, the fits used sizes or masses that were up to 40% passing only. He found that there was insignificant difference between the three distribution fits. As might be expected, two-component distributions provided better fits than did single-component distributions, but they were more sensitive to starting guesses used in the non-linear fitting process.

The scaled CDF of the sigmoidal function is defined as,

$$F(x)=100\left(\frac{(x/a)^b}{1+(x/a)^b}\right) \tag{10}$$

and yields a median size of $x = a$ and a mean size of $\pi a/(b \sin(\pi/b))$ when $a > 0$ and $b > 1$. Note that for $b < 1$ the PDF of the sigmoidal function has a singularity and the mean is not defined.

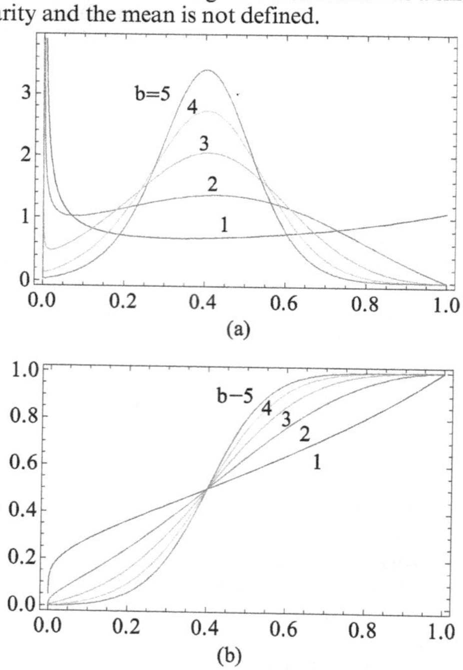

Figure 5. A series of Swebrec functions (a) PDFs shown as normalised frequency distributions and (b) associated CDFs. The distributions in this case are for normalised Swebrec functions with $x_{max} = 1$, $x_{50} = 0.4$, and b ranging from 1 to 5.

Figure 6 shows an example of the PDF for the sigmoidal function and its CDF.

1.5 *Three-parameter modified Weibull model*

The three-parameter Swebrec function fits a wide range of size distributions from blasting, crushing and grinding. Its ability to do this appears to be due, in part, to the high fraction of finer particles for a range of its parameters. Figure 5 shows this behaviour for values of the undulation parameter, $b > 1$.

The traditional three-parameter Weibull distribution introduces a location parameter that shifts the whole distribution to lower sizes so that the range of the independent variable (size) is from minus infinity to plus infinity (Evans et al., 2000). Clearly, this is not satisfactory for the present application as we cannot have negative sizes.

An alternative ad-hoc approach is taken here that modifies the Weibull distribution but retains only positive sizes. An extra parameter is defined as a fines size parameter, *sz* and the new three-parameter CDF is formed from a prototype two-parameter Weibull using the following procedure. Produce the two-parameter Weibull PDF and consider the area up to the size, *sz*. This is shown

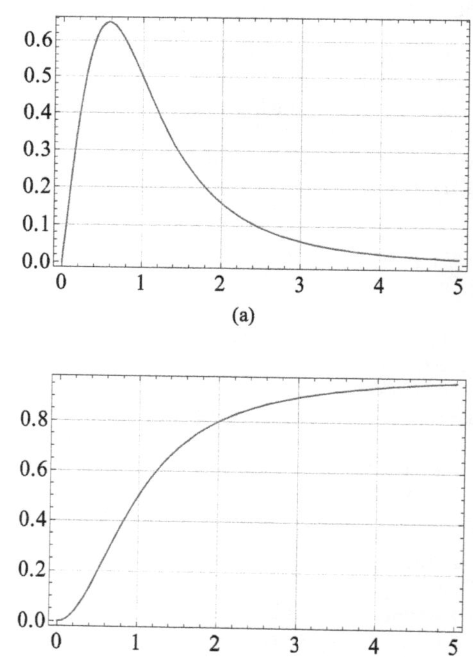

Figure 6. (a) PDF of the sigmoidal model and (b) its associated CDF. The distribution parameters are $a = 1$ and $b = 2$.

as the shaded area in Figure 7 for the particular case of $x_c = 0.4$, $n = 2$ and $sz = 0.1$. Reflect the area about $sz / 2$ and add that area to the original PDF. Of course, the total area under the PDF must equal unity (see Table 2) so we must scale the PDF appropriately to meet this requirement. The prototype PDF and its three-parameter modified form are shown in Figure 8.

The actual equation for the PDF of this three-parameter modified Weibull distribution is,

$$
\begin{aligned}
f(x) &= n\frac{(sz-x)^{n-1}}{x_c^n}\,\mathrm{Exp}\left(\left(\frac{sz-x}{x_c}\right)^n\right) \\
&\quad + n\frac{x^{n-1}}{x_c^n\left(2-\exp\left(-\left(\frac{sz}{x_c}\right)^n\right)\right)}\exp\left(-\left(\frac{x}{x_c}\right)^n\right) \qquad 0 \le x \le sz \\
F(x) &= n\frac{x^{n-1}}{x_c^n\left(2-\exp\left(-\left(\frac{sz}{x_c}\right)^n\right)\right)}\exp\left(-\left(\frac{x}{x_c}\right)^n\right) \qquad x \ge sz
\end{aligned}
\tag{11}
$$

1.6 *Combined exponential/Weibull model*

The three-parameter modified Weibull CDF described above is a piecewise combination with generally discontinuous behaviour in the derivative at sz. Combining an exponential and Weibull distribution can form an alternative three-parameter CDF that emphasizes the fines region.

The CDF for a combined exponential/Weibull function is given by,

$$
F(x) = \left(\frac{1}{2}\right)\left[2-\exp\left(\frac{-x}{b}\right)-\exp\left(-\left(\frac{x}{x_c}\right)^n\right)\right] \tag{12}
$$

where b is the decay for the exponential function and the characteristic size, x_c, and the uniformity index, n are the usual Weibull function parameters. Figure 9 shows two examples of this function.

1.7 *Combined Rayleigh/Weibull model*

Combining a Rayleigh and Weibull distribution can form another three-parameter CDF that also emphasizes the fines region.

The CDF for a combined Rayleigh/Weibull function is given by,

$$
F(x) = \left(\frac{1}{2}\right)\left[2-\exp\left(\frac{-x^2}{2b^2}\right)-\exp\left(-\left(\frac{x}{x_c}\right)^n\right)\right] \tag{13}
$$

where b is the decay for the Rayleigh function and the characteristic size, x_c, and the uniformity index, n are the usual Weibull function parameters. Figure 10 shows an example of this function.

2 FITTING THE CDF TO DATA

In order to assess the relevance of any size distribution function, we require some data to which we fit the function. It is customary to try and fit the CDF to such data and that is the procedure used here. Furthermore, given the statement by Ouchterlony (2005) that "The new Swebrec© function gives excellent fits to hundreds of sieved fragmentation data ...", it seems natural to generate data that is produced by a Swebrec function for the purposes of simulation. For comparison, data that follows a Rosin-Rammler distribution is also created. Measured data from Ouchterlony (2005) and from Blair (2004) are used also to examine a range of fitting functions.

The FindFit (or equivalently the Nonlinear-ModelFit) function within the software package Mathematica© (version 7.0.1.0) (Wolfram, 1999) was used to conduct the fits. It accesses a variety of methods to perform the non-linear fit, including the Levenberg-Marquardt method. As with all

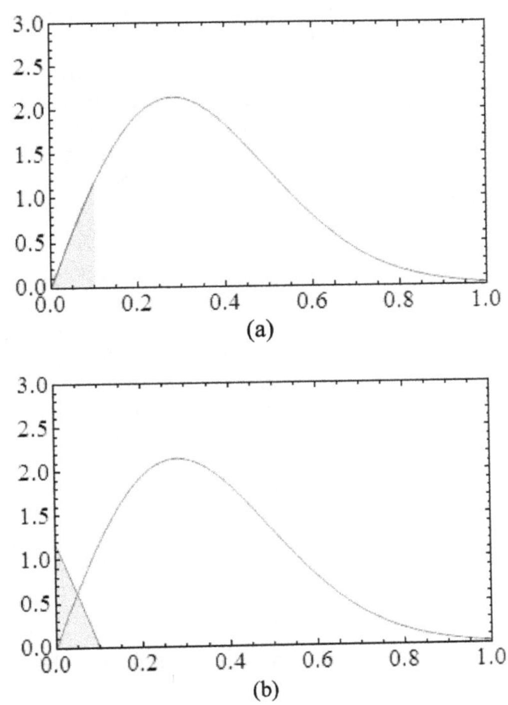

Figure 7. Two-parameter Weibull PDF ($x_c = 0.4$, $n = 2$) with $sz = 0.1$. (a) the shaded area of the PDF that is reflected about $sz / 2$ (b) the reflected area is shown.

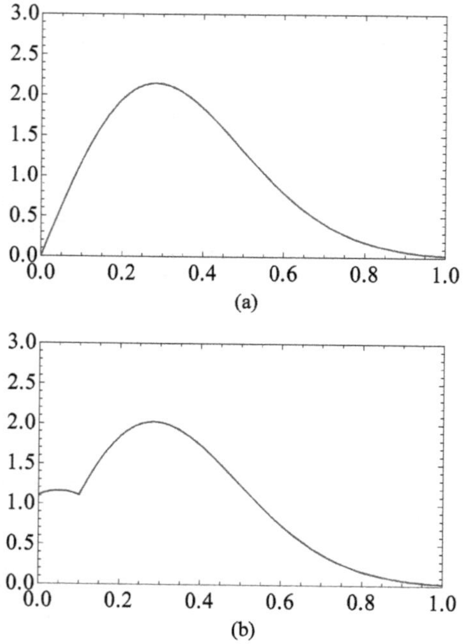

Figure 8. (a) Prototype two-parameter Weibull PDF ($x_c = 0.4$, $n = 2$) (b) the three-parameter modified Weibull PDF on the right ($sz = 0.1$). The smaller sizes are emphasised in the resultant distribution.

such non-linear fitting procedures, the robustness of the fit depends on the starting values, the function being fit to the data and the data. A measure of the success of the fit is the sum of the residuals between the data and the fitting function. A global minimum in the residuals is not guaranteed when conducting such non-linear fitting.

2.1 *Test data*

A test data set was created using the Swebrec function Equation (8) with the following parameters: x_{max} = 1000 mm, x_{50} = 400 mm and b = 2. A second data set was created using the Rosin-Rammler function (Equation (1)) with the parameters x_c = 400 mm and n = 2. A series of other data sets were generated by digitising data from several figures in Ouchterlony (2005). Finally, data from Blair (2004) is also used for testing the CDF fits. The test data sets are summarised in Table 3.

2.2 *The CDF function fits*

The following CDFs have been used to fit the test data sets:

- two–parameter Rosin-Rammler or Weibull function
- three–parameter Swebrec function
- sigmoidal function

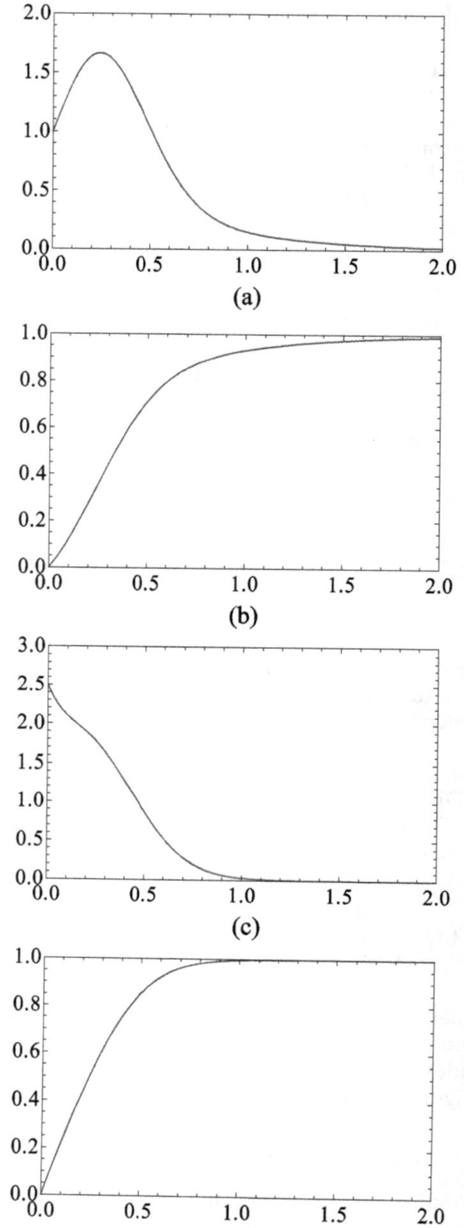

Figure 9. (a) PDF and (b) CDF (upper for a combined exponential/Weibull function with $b = 0.5$, $x_c = 0.4$ and $n = 2$. (c) PDF and (d) CDF for a combined exponential/Weibull function with $b = 0.2$, $x_c = 0.4$ and $n = 2$.

- a modified three–parameter Weibull function
- combined exponential/two–parameter Weibull function
- combined Rayleigh/two–parameter Weibull function.

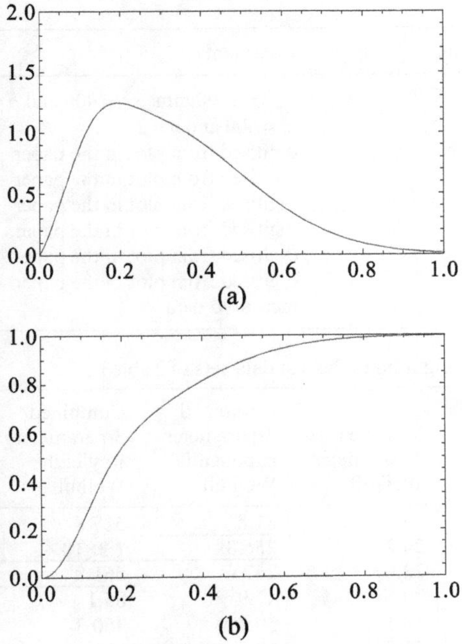

Figure 10. (a) PDF and (b) CDF for a combined Rayleigh/Weibull function with $b = 0.1$, $x_c = 0.4$ and $n = 2$.

An example of the starting parameters for a fit, the best fit parameters for the fit, plots of the fits and plots of the residuals are given in the Appendix. The residual plots tend to show some structure, indicating that the fits are not capturing some apsects of the data. Ouchterlony (personal communication, 2010) indicated that he often uses a weighted non-linear fit in his parameter estimates. The weighting is the inverse of the square-root of the size. Such weighting emphasizes the smaller sizes in the distribution and promotes a better fit in that region of the distribution. The fits documented in the Appendix use both uniform weighting as suggested by Ouchterlony. A summary of the normalised residuals for the uniform weighting fits is given in Table 4. Table 5 contains the summary of normalised residuals for non-uniform weighting fits. The normalised residuals are calculated as the sum of the squared differences between the actual data and the fitted curve at the same sizes. This measure emphasises the differences between the data and the fits compared to the square root of the sum of the squared residuals as the fitting metric. The residuals have also been scaled for a range of 0–100 in the first two data sets so the results may be compared directly to the other data sets that have a range between 0–1.

The fits in Tables 4 and 5 have been highlighted as dark grey (minimum residual), medium grey (next to minimum residual and light grey (second to minimum residual). It is clear that of all the CDF fits, the Swebrec function fits most of the data superbly. This reinforces the statement by Ouchterlony (2005) that "The new Swebrec function gives excellent fits to hundreds of sets of sieved fragmentation data …".

The next best fits in Tables 4 and 5 are the modified 3-parameter Weibull fit and the combined exponential/Weibull fit. Of these two, the combined exponential/Weibull fit provides a somewhat better overall fit to the data than does the modified 3-parameter Weibull fit. The latter CDF is somewhat ad hoc while both CDFs have the Weibull fit as an integral component function. It is speculated that the combined exponential/Weibull CDF is a more natural description and may have a mechanistic explanation. This is described later and while there are caveats at this stage, the concept deserves further investigation. The poorest overall fit to the data is the sigmoidal fit as given in Tables 4 and 5. The data in Tables 4 and 5 have been plotted in Figure 11.

3 MECHANISMS AND DISTRIBUTION FUNCTIONS

From a blasting engineer's perspective, it is useful to design a blast and be able to predict the resulting rock fragmentation distribution since this influences the efficiency of digging machinery and downstream processes such as crushing and grinding. It is also useful to know how alterations in the geometry, rock type, explosive type, and initiation sequence affect the resulting rock fragmentation distribution. The Kuz-Ram model offers a solution that excludes one of these elements—blast timing—but the parameters in the model are defined by quantities that are measurable in the field.

It can be useful to summarise measurements using various curve fits—indeed that has been done above—but more appealing is to fit a curve that has some basis in the physical processes causing the fragments creation. A combined exponential-Weibull distribution that has just three parameters has been suggested as a possible candidate for fitting rock fragmentation distributions. It is worth noting that the Rosin-Rammler or Weibull distribution has some foundation in a mechanistic or physical principle based on sequential fragmentation by a branching tree of cracks (Brown and Wohletz, 1995). While this is a somewhat phenomenological approach rather than a strict micromechanical approach, it at least bestows a physical basis that may explain why the distribution may fit measured data reasonably well. Interestingly, Brown and Wohletz (1995) also suggest that the suitability of a lognormal distribution for such fits may be fortuitous.

Table 3. Summary of test data sets for CDF fits.

Test data	Number of points	Comments
1. Swebrec CDF	35	x_{max} = 1000 mm, x_{50} = 400 and b = 2
2. Rosin-Rammler CDF	35	x_c = 400 and n = 2
3. Ouchterlony (2005) Figure 1	24	digitised from plot in the paper
4. Ouchterlony (2005) Figure 9	20	digitised from plot in the paper
5. Ouchterlony (2005) Figure 14	16	digitised from plot in the paper
6. Ouchterlony (2005) Figure 15	18	digitised from plot in the paper
7. Ouchterlony (2005) Figure 16	12	digitised from plot in the paper
8. Ouchterlony (2005) Figure 17	7	digitised from plot in the paper
9. Blair (2004) Figure 1 – Set 1	19	measured data

Table 4. Normalised residuals from CDF fits using uniform weighting to the test data sets of Table3.

Test data set	3-parameter Swebrec	2-parameter Weibull	Sigmoidal	Modified 3-parameter Weibull	Combined 3-parameter exponential/ Weibull	Combined 3-parameter Rayleigh/ Weibull
1	1.4×10^{-19}	419.7	836.5	97.6	81.8	517.8
2	7.9	1.7×10^{-27}	119.3	24.2	256.3	1.3×10^{-27}
3	84.8	77.7	261.3	223.7	262.6	561.8
4	10.8	182.4	371.2	70.3	47.9	85.1
5	33.5	481.8	844.3	58.1	59.9	360.3
6	68.7	97.9	225.5	79.8	82.5	219.7
7	27.1	439.8	696.3	330.3	326.0	73.7
8	6.3	24.3	134.9	6.1	23.6	30.2
9	3.8	12.0	22.2	5.1	3.0	5.7

Table 5. Normalised residuals from CDF fits using inverse square-root weighting to the test data sets of Table 3.

Test data set	3-parameter Swebrec	2-parameter Weibull	Sigmoidal	Modified 3-parameter Weibull	Combined 3-parameter exponential/ Weibull	Combined 3-parameter Rayleigh/ Weibull
1	1.7×10^{-25}	667.3	1188.8	345.2	94.2	399.7
2	10.2	1.4×10^{-26}	130.5	31.8	328.8	1.1×10^{-27}
3	97.9	81.5	438.3	1382.9	192.1	652.0
4	13.8	379.8	622.1	107.6	56.0	214.0
5	38.9	656.1	1077.3	75.6	69.9	484.4
6	70.0	157.8	455.0	107.8	106.7	162.6
7	36.6	503.5	696.3	359.5	346.5	75.3
8	6.7	32.0	189.2	6.5	31.0	31.3
9	9.5	18.1	36.4	5.10	5.06	5.8

A Weibull distribution reduces to an exponential distribution when the uniformity index is unity (see Equation (1)); so it may be argued that the combined exponential-Weibull distribution also has a physical basis as it is the combination of two separate Weibull distributions, one of which is simplified. Furthermore, an exponential distribution (Evans et al., 2000) arises from a Poisson process of fracture points (Grady and Kipp, 1985) that offers an alternative mechanism for the generation of fragments biased towards the fines end in the combined exponential-Weibull distribution. Interestingly, these two mechanisms, namely a Poisson process of fracture points and some form of crack nucleation and branching, appear to have been used to explain the opposite ends of a cumulative distribution function to that used here (Ouchterlony and Moser, 2006).

4 DISCUSSION

The Kuz-Ram model (Cunningham, 1983, 1987) with the correction by Spathis (2004) may be used to estimate the fragment size distribution produced by blasting. The accuracy of this prediction is not

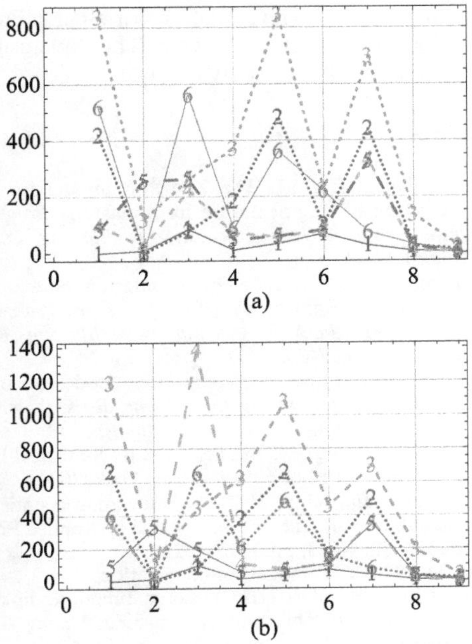

Figure 11. (a) Plot of the residuals for uniform weighted fits (b) plot of the residuals for non-uniform weighted fits. The fits are: 1 = Swebrec, 2 = Weibull, 3 = sigmoidal, 4 = modified Weibull, 5 = combined exponential/Weibull and 6 = combined Rayleigh/Weibull.

assessed here although some discussion on the sensitivity of its two paremeters has been provided. Given that the Kuz-Ram model as used in blasting is well over twenty years old, it is not surprising that some improvements to it have been sought. Researchers have focused on the relatively poor estimate of the finer size fractions by the Kuz-Ram model (Djordjevic, 1999; Thornton, 2001). The Swebrec CDF discovered by Ouchterlony (2005) provides an alternative to the Weibull CDF as used in the Kuz-Ram model. The Swebrec CDF provides the best fit of any of the two-parameter or three-parameter CDF models used here and this further substantiates the demonstrations of its efficacy in a wide range of fragmentation measurements as given by Ouchterlony and his co-workers.

A key issue that has arisen in the present study is the incorrect equating of the mean size to the median (or 50% passing) size, previously noted by Spathis (2004) in relation to the original Kuz-Ram model. This is important because in using the Kuz-Ram model, the estimate for the mean size is provided by Kuznetsov (1973). That estimate is equated to the mean size of the Weibull distribution as given in Equation (2).

Now when Ouchterlony (2005) maintains the Kuznetsov estimate for the mean size but replaces the Rosin-Rammler or Weibull CDF by the Swebrec function, he also suggests equating the Kuznetsov mean size to the median size of the Swebrec PDF. A reason for doing this is that apparently there is no analytical expression for the mean size of the Swebrec PDF. This creates a fundamental issue for the straight-forward implementation suggested by Ouchterlony (2005) of using the Swebrec CDF in a Kuz-Ram style model. So, how can one proceed?

Several opportunites arise for involving the Swebrec function in a Kuz-Ram approach:

- ignore the formal difference between the mean and median sizes and use the process suggested by Ouchterlony. This is not a preferred approach.
- for cases where the calculated uniformity index (Equation 2) is greater than 2, proceed with the process suggested by Ouchterlony—in this case we have less than approximately 5% difference between the mean and median size for a Weibull distribution. This approach is acceptable for more uniform rock size distributions.
- use Cunningham's approach to determine the two parameters for the Rosin-Rammler (Weibull) curve and calculate the 50% passing size from the curve fit parameters (Equations (2) and (3)), estimate the maximum size using the maximum joint spacing or the spacing or burden for the blast, and use Equation (9) to estimate the undulation parameter. Use the Swebrec function to predict the rock fragmentation distribution.

The last approach above takes most of what is recommended by Ouchterlony (2005) and circumvents the formally incorrect assignment of the Kuznetsov mean size to the median (50% passing) size. The Appendix examines the approach when we have measured data and compares it to a direct non-linear Swebrec fit to the data. In almost all cases the fit using the indirect approach is inferior to the direct non-linear Swebrec fit.

Ouchterlony (2005a, 2009) has described a different approach to predicting the rock fragmentation distribution. It relies on equating the slope of the Swebrec function at the median size to the slope found from experimental data for full-scale blasts. This provides an equation for the undulation parameter in terms of the median size and thus reduces the Swebrec fit to a two parameter model. The evaluation of the slope of any function is an inherently noisy process and prone to error. Indeed the best way to contain errors in its determination is to fit a smooth curve to data and differentiate the smooth curve. As shown in the Appendix, two different Swebrec curve fits to data arise when one assumes a given slope at the median size versus when one allows all three parameters to be free in a curve fit.

It is important to remember that methods for the prediction of the fragmentation size distribution of rock produced by blasting are there to assist blasting engineers design blasts that produce desired outcomes. An essential component of this process is to take mesurements and ensure that such predictions accord with observed behaviour.

5 CONCLUSIONS

A three parameter rock fragmentation distribution function is proposed that fits sets of idealized and measured data almost as well as the Swebrec function. The new function combines an exponential and Weibull distribution and retains the strength of the often-used Rosin-Rammler (Weibull) model while adding the benefit of fitting finer sizes better than does the Weibull distribution alone. Furthermore, the new model can be incorporated readily into a modified Kuz-Ram model where the Rosin-Rammler curve is replaced by the new three-parameter model. This may be done without some of the formal difficulties identified when using the Swebrec function in that way. A further strength of the new model is that its component distributions have some foundation in physical principles or mechanisms that have been suggested apply in blasting rock. Caution is required when using alternative methods of fitting the Swebrec function, whether it be by equating the slope of the Weibull fit at the median size to the slope of the Swebrec function there, or when using measured experimental data for the slopes at the median size.

The following recommendations are made:

- The three parameter rock fragmentation distribution function proposed here should be applied to other fragmentation data and compared to the remarkable fits that occur using the Swebrec function.
- The long-standing work of Cunningham uses explosives, rock mass properties and blast design geometry to determine the parameters of the Rosin-Rammler curve. It would be worth investigating if similar approaches can be developed to determine the three parameters in the new function which will assist blasting engineers better predict the rock fragmentation produced by blasting.

ACKNOWLEDGEMENTS

I would like to acknowledge discussions with Finn Ouchterlony, Alan Minchinton, and Michael Noy. I acknowledge the review of the original submission by an anonymous referee. The conclusions contained in this report are my own.

REFERENCES

Cunningham, C. 1983. The Kuz-Ram model for the prediction of fragmentation from blasting. *Proc. 1st Intnl. Symp. Rock Fragmentation by Blasting, Lulea, August 22–26.*

Cunningham, C.V.B. 1987. Fragmentation estimations and the Kuz-Ram model—four years on. *Proc. 2nd Intnl. Symp. Rock Fragmentation by Blasting, Keystone, August 23–26.*

Djordjevic, N. 1999. Two-component model of blast fragmentation. *Proc. 6th Intnl. Symp. Rock Fragmentation by Blasting, Johannesburg, August 8–12.*

Evans, M., Hastings, N. and Peacock, B.: Statistical Distributions, 3rd edition. Wiley, New York, 2000.

Grady, D.E. and Kipp, M.E. 1985: Geometric statistics and dynamic fragmentation, Journal of Applied Physics, Vol. 58, No. 3, pp. 1210–1222.

Noy, M.J. (2012). Personal communication.

Ouchterlony, F. 2005. The Swebrec function: linking fragmentation by blasting and crushing. *Mining Technology* (*Trans. Inst. Min. Metall. A*), 114, A1–A16.

Ouchterlony, F. 2005a. What does the fragment size distribution of blasted rock look like? Proc. 3rd EFEE World Conference on Explosives and Blasting, Brighton.

Ouchterlony, F. and Moser. P. 2006. Likenesses and differences in the fragmentation of full-scale and model-scale blasts. *Proc. 8th Intnl. Symp. Rock Fragmentation by Blasting, Santiago, May 7–11.*

Ouchterlony, F. 2009. A common form for fragment size distributions from blasting and a derivation of a generalized Kuzntetsov's x_{50}-equation. *Proc. 9th Intnl. Symp. Rock Fragmentation by Blasting, Granada, September 13–17.*

Sanchidrian, J.A, Segarra, P., Lopez, L.M., Ouchterlony, F., and Moser, P. 2009. Evaluation of some distribution functions for describing rock fragmentation data. *Proc. 9th Intnl. Symp. Rock Fragmentation by Blasting, Granada, September 13–17.*

Singh, S.P., VanDoorselaere, D. and Narendrula. R. 2006. Comparison of sieved, predicted and measured blast fragmentation results. *Proc.32nd Ann. Conf. On Expl. And Blasting Techniques, Dallas, Jan. 29 –Feb. 1.*

Spathis, A.T. 2004. A correction relating to the analysis of the original Kuz-Ram model. *Int. J. Blasting and Fragmentation*, 8, No.4, 201–205.

Spathis, A.T. 2009. Formulae and techniques for assessing features of blast-induced fragmentation distributions. *Proc. 9th Intnl. Symp. Rock Fragmentation by Blasting, Granada, September 13–17.*

Thornton, D, Kanchibotla, S.S. and Brunton, I. 2001. Modelling the impact of rockmass and blast design variation on blast fragmentation. *Proc. EXPLO 2001, Hunter Valley, October 28–31, Melbourne: AusIMM.*

Wolfram, S. 1999. The Mathematica Book (4th Edition). Cambridge University Press, Cambridge.

APPENDIX

Figures A1 and A2 give an example fit of the exponential/two-parameter Weibull function to Data Set 6 from Table 3. Reference to Tables 4 and 5 shows that it performs quite well with and without weighting at the lower sizes. The data spans over three decades of fragment sizes.

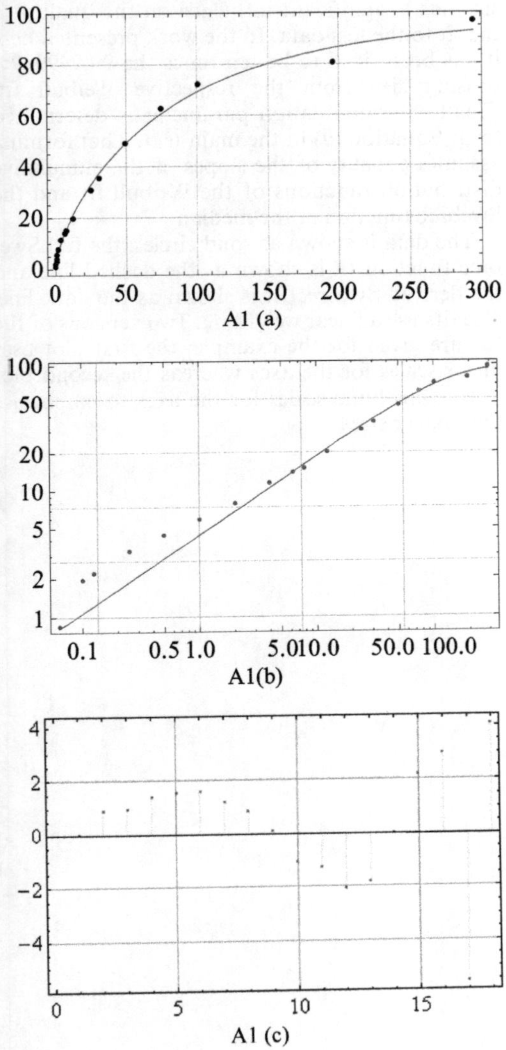

Figure A1. Combined exponential/two-parameter Weibull fit to Data Set 6 without weighting (a) linear scales (b) logarithmic scales (c) residuals. Starting values: {b, 100}, {xc, 50}, {n, 2}. Best fit parameters are {b = 82.9818, xc = 78.3395, n = 0.58336}.

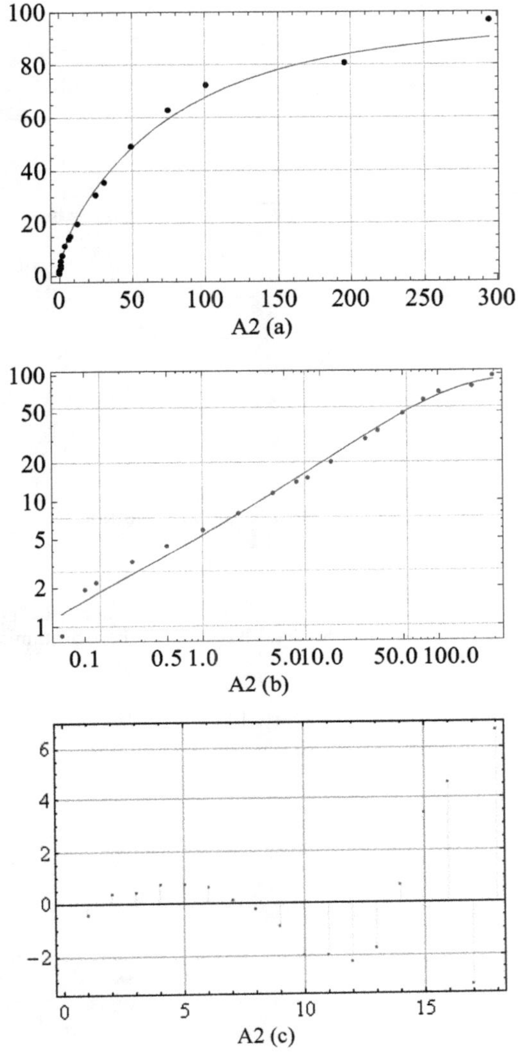

Figure A2. Combined exponential/two-parameter Weibull fit to Data Set 6 with inverse square-root weighting (a) linear scales (b) logarithmic scales (c) residuals. Starting values: {b, 100}, {xc, 50}, {n, 2}. Best fit parameters are {b = 80.6255, xc = 95.1184, n = 0.506529}.

Figure A3 is an example of two different methods used to fit the test data. The first is a standard non-linear least squares (NLS) fit and that has been compared with other CDF fits of various functions in the main text with an example given here.

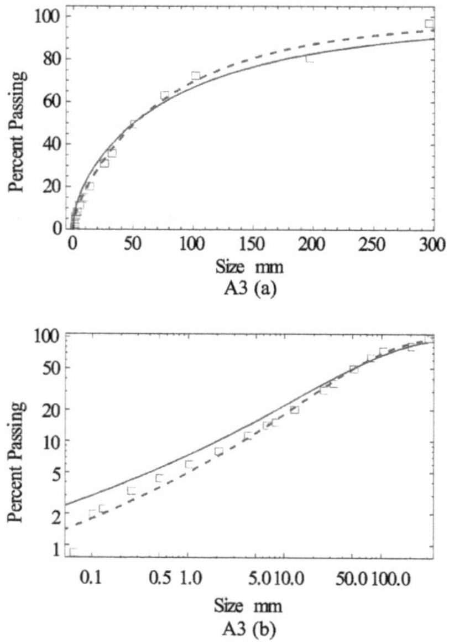

Figure A3. NLS fit and derived fit based on Rosin-Rammler for Data Set 6 (a) linear scales (b) logarithmic scales.

The second fit is based on first using a Rosin-Rammler (Weibull) fit to the data that yields the two parameters of such a fit: The characteristic size and the uniformity index. These two parameters are used to derive the median (50% passing) size for the fit using Equation (3) in the main text. The other two parameters for the Swebrec function are the maximum size and the undulation parameter. It has been found that the maximum size has a significant influence on the quality of the fit to the test data. In the work presented here it has been fixed to be ten times the 99.999999% passing size from the respective Weibull fit. Finally, the undulation parameter is determined from Equation (9) in the main text. That formula assumes equality of the slopes of the cumulative distribution functions of the Weibull fit and the Swebrec function at the median.

The data is shown as solid circles, the full Swebrec function fit is shown as the dashed line and the derived Swebrec fit is shown as the dark line. The fits use a linear weighting. Two versions of the fits are given for the example: the first plot uses linear scales for the axes whereas the second plot uses logarithmic scales for the axes to emphasise the smaller sizes.

Measurement and Analysis of Blast Fragmentation – Sanchidrián & Singh (Eds)
 ISBN 978-0-415-62140-3

On the performance of truncated distributions to describe rock fragmentation

J.A. Sanchidrián, P. Segarra & L.M. López
Universidad Politécnica de Madrid—ETSI Minas, Spain

F. Ouchterlony & P. Moser
Montanuniversität Leoben, Austria

ABSTRACT: Weibull (Rosin-Rammler), Grady, lognormal, log-logistic and Gilvarry distributions are compared with their variants truncated at a maximum size. The Swebrec distribution (a truncated one) is included in the comparison. The comparison is made on a benchmark of 448 sets of screened fragment size data from blasted and crushed rock of different origins. The truncated distributions reduce the unexplained variance by a factor of more than five when compared with the infinite variable ones, except the Gilvarry. The higher determination coefficients are obtained by the truncated Weibull and the Swebrec, with median values 0.9975 and 0.9974, respectively. The errors in predicting sizes have been determined for each of the distribution functions across the different passing ranges. Relative errors in size determination are 6 to 7% in the coarse zone (>80% passing) for all truncated functions except Gilvarry; similar values are obtained in the central range (20 to 80% passing) for the truncated Weibull, Swebrec and Grady. These three distributions are also the best ones in the fine zone (2 to 20% passing) though relative errors are here 25 to 30%. Errors in the very fine range (<2%) are generally wild, with the best median value of 180%, for the truncated Weibull. When analyzed for the different materials, truncated Weibull and Swebrec are consistently the distributions obtaining lower size estimation errors in the various zones, with a noticeable generally good behavior of the truncated log-logistic in the coarse range, and to a lesser extent also in this range the truncated lognormal and Grady.

1 INTRODUCTION

The assessment of fragmentation in blasting and in any of the subsequent crushing and grinding stages is an important issue in mining control and optimization. Fragmentation characteristics influence the mucking productivity, the crusher throughput and energy consumption, the plant efficiency, yield and recovery, and the price itself of the end product in the case of industrial minerals and aggregates.

Fragment size distributions are a statistical representation of the population of fragment, or particle, sizes. In the context of rock fragmentation, size is defined with respect to the smaller square mesh size through which the particle can pass (King 2001, Gupta & Yan 2006). The size distribution is typically expressed as cumulative probability in the form of a cumulative distribution function, giving the probability that a fragment be smaller than a given size. Such probability may be estimated from histograms of numbers of particles (as image analysis systems do) or from histograms of mass fractions in the common industrial sieving way.

Rock fragmentation has been described for many years, almost exclusively, by means of the Rosin-Rammler distribution; in the last seven years, the Swebrec function has gained a relatively high profile as it has been shown to represent the fragmented rock sizes with advantage to the Rosin-Rammler both in the fines and in the coarse ends. Sanchidrián et al. (2012) have recently compared the performance of the following distribution functions: Weibull-Rosin-Rammler, Grady, lognormal (all two-parameter distributions), the three-parameter Gilvarry, and their double-linear combinations in the form of bi-component or bi-modal distributions. Together with these, the three-parameter, truncated Swebrec function, and the extended Swebrec, a five-parameter, truncated, bi-component function were also analyzed. These functions were fitted to 448 sets of screened fragment size data from blasted and crushed rock of a variety of origins. The errors in predicting sizes were determined for each of the distribution functions across the whole range of data, dividing the passing range in four zones, coarse (>80%), central (80–20%), fine (20–2%) and very fine (<2%). Swebrec was by far the best single component function in all zones, comparable to the best bi-components in the coarse and central, with relative errors in size determination of 6 to 7% in relative value.

Indeed, it is not surprising that a three-parameter function such as the Swebrec outdoes two-parameter ones like the Weibull-Rosin-Rammler, Grady and lognormal, and raises the question of how these functions would behave if a truncated form of them – similar to the Swebrec, including as third parameter the truncation size, that meets the condition of an accumulated value of 1 (or 100%) – were employed. The present work addresses this point by comparing functions with two parameters (the already mentioned Rosin-Rammler, Grady and lognormal, to which the log-logistic has been added) with their truncated variants, and with the Swebrec function itself. The exponential, three-parameter Gilvarry distribution and a four-parameter truncated version of it, has also been included in the analysis. The comparison is made on the same benchmark of 448 data sets mentioned above.

2 DISTRIBUTION FUNCTIONS

The representation of the sizes of a granular material is generally done in the form of a cumulative probability, or cumulative distribution function (CDF), F, giving the probability p that a fragment be smaller than the variable x (i.e. p is the fraction passing a mesh size x):

$$P(x) = F(x, \bar{\pi}) \tag{1}$$

$\bar{\pi}$ being the vector of parameters of the function. The functions that have been studied in the present work are the following (in parenthesis the abbreviations used for conciseness):

– Weibull-Rosin-Rammler (WRR; Rosin & Rammler 1936, Weibull 1939, 1951):

$$F_{WRR} = 1 - \exp\left[-\left(\frac{x}{x_c}\right)^n\right], \quad 0 \le x \le \infty \tag{2}$$

where x_c and n are the scale and shape parameters, respectively.

– Grady (GRA, Grady & Kipp 1985, Grady 1990):

$$F_{GRA} = 1 - \left[1 + \left(\frac{x}{x_g}\right)^\alpha\right] \exp\left[-\left(\frac{x}{x_g}\right)^\alpha\right], \quad 0 \le x \le \infty \tag{3}$$

where x_g and α are the scale and shape parameters, respectively.

– Lognormal (LGN):

$$F_{LGN} = \frac{1}{s\sqrt{2\pi}} \int_0^x \frac{1}{t} \exp\left[-\frac{(\ln t - x_m)^2}{2s^2}\right] dt, \quad 0 \le x \le \infty \tag{4}$$

where x_m and s are the location and scale parameters (the mean and the standard deviation of the natural logarithm of x).

– Log-logistic (LGL):

$$F_{LGL} = \frac{1}{1 + \left(\frac{x}{x_{50}}\right)^{-\gamma}}, \quad 0 \le x \le \infty \tag{5}$$

where x_{50} (the median) and γ are the scale and shape parameters, respectively.

– Gilvarry (GIL; Gilvarry 1961, Gilvarry & Bergstrom 1961, Sil'vestrov 2004 a, b):

$$F_{GIL} = 1 - \exp\left\{-\left[\left(\frac{x}{x_1}\right) + \left(\frac{x}{x_2}\right)^2 + \left(\frac{x}{x_3}\right)^3\right]\right\}, \quad 0 \le x \le \infty \tag{6}$$

where x_1, x_2 and x_3 are first, second and third order scale parameters.

The truncated versions of these are derived by substituting ξ for x:

$$\xi = \frac{x}{x_{max} - x} \tag{7}$$

Leading to (e.g. for WRR):

$$F_{TWRR} = 1 - \exp\left[-\left(\frac{x}{x_c}\frac{x_{max} - x_c}{x_{max} - x}\right)^n\right], \quad 0 \le x \le x_{max} \tag{8}$$

And similar expressions for the other distributions, meeting the condition $F = 1$ at $x = x_{max}$. Truncated functions are referred here with the normal, infinite variable, function acronym preceded by a T.

Finally, the Swebrec (SWE, Ouchterlony 2005a, b) cumulative distribution is:

$$F_{SWE} = \frac{1}{1 + \left[\frac{\ln(x_{max}/x)}{\ln(x_{max}/x_{50})}\right]^b}, \quad 0 \le x \le x_{max} \tag{9}$$

where x_{50} is the median and b a shape parameter.

3 FRAGMENTATION DATA AND DISTRIBUTION FITTING

The fragment size data used is described in detail by Sanchidrián et al. (2012). It comprises 448 curves, or data sets, spanning a wide range of sizes and with variable characteristics and origins; they are all obtained by sieving. Tables 1a–b give a summary of them. For each data source, the range of the number of data points of each data set, n_p, the ranges of minimum and maximum sizes, x_{min} and x'_{max} (the prime is used to differentiate the maximum size value of data set from the truncation size, x_{max}, a parameter to be determined from the fit) and the ranges of the minimum percentage passing are given, together with the following dimensionless ratios:

– Log size range: $r_x = \log_{10}(x'_{max} / x_{min})$
– Log passing range: $r_p = \log_{10}(p_{max} / p_{min})$
– Overall log slope: $s_L = r_p / r_x$

Values given in Table 1 are the extremes of these for each referenced source of data sets.

The eleven distributions: WRR, GRA, LGN, LGL, GIL, TWRR, TGRA, TLGN, TLGL, TGIL and SWE have been fitted to the 448 data sets using an ordinary least squares scheme. The

Table 1a. Fragmentation data: Blasted material.

Mine or quarry site, or rock origin	Rock	No. of sets	n_p	x_{min} (mm)	x'_{max} (mm)	p_{min} (%)	r_x	r_p	s_L
1 Blasted, mine (78 data sets)									
El Alto, Spain[a]	Limestone	1	16	0.063	800	0.36	4.1	2.4	0.6
Christmas mine, USA[b]	Copper ore	3	10	19.05	381	12.1–19.9	1.3	0.7–0.9	0.5–0.7
Mt. Coot-tha, Australia[c]	Hornfel	1	24	0.355	2000	2.75	3.8	1.6	0.4
Bårarp, Sweden[d]	Granite	7	19–20	0.075	500–1000	0.1–0.6	3.8–4.1	2.2–2.9	0.5–0.7
Kållered, Sweden[e]	Gneiss	6	17	0.075	2000	0.34–0.49	4.4	2.3–2.5	0.5–0.6
Billingsryd, Sweden[e]	Dolerite	6	18	0.074	2000	0.76–1.03	4.4	2.0–2.1	0.4–0.5
South Africa[f]	Gold reef	8	8–26	0.053–0.075	300–304	0.16–3.1	3.6–3.8	1.5–2.8	0.4–0.8
South Africa[f]	Granite	17	7	10	500	4.7–12.5	1.7	0.9–1.3	0.5–0.8
Rolla, USA[g]	Dolomite	29	6–8	9.525	229–457	7.9–20.6	1.4–1.7	0.7–1.1	0.5–0.7
2 Blasted, specimens (197 data sets)									
Not reported[h]	Limestone	1	15	0.002	203	0.0041	5.0	4.4	0.9
Darley Dale, UK[h]	Sandstone	12	16	0.002	406	0.0035–0.029	5.3	3.5–4.5	0.7–0.8
Imberg, Germany[i]	Sandstone	3	19–21	0.063	80–125	0.13–0.45	3.1–3.3	2.3–2.8	0.7–0.8
Eibenstein, Austria[j]	Amphibolite	15	14–20	0.063	25–125	0.14–3.55	2.6–3.3	1.4–2.8	0.5–0.9
Bårarp, Sweden[d,k]	Granite	7	15–21	0.063	31.5–125	0.21–3.20	2.7–3.3	1.5–2.7	0.6–0.8
El Alto, Spain[j]	Limestone	12	14–21	0.063	25–125	0.13–4.00	2.6–3.3	1.4–2.9	0.5–0.9
Klinthagen, Sweden[j]	Limestone-stroma[1]	9	15–20	0.063	31.5–100	0.18–1.86	2.7–3.2	1.7–2.7	0.6–0.9
Klinthagen, Sweden[j]	Limestone-crin[1]	9	13–21	0.063	20–125	0.16–1.89	2.5–3.3	1.7–2.8	0.6–0.9
Klinthagen, Sweden[j]	Limestone-mass[1]	9	14–21	0.063	25–125	0.20–2.66	2.6–3.3	1.6–2.7	0.6–0.9
Klinthagen, Sweden[j]	Limestone-frag[1]	10	13–20	0.063	20–100	0.32–3.41	2.5–3.2	1.5–2.5	0.6–0.8
Norway[l,m]	Syenite, granite, gabbro, gneiss	21	9–11	0.25–0.50	100–150	0.06–1.85	2.3–2.8	1.7–3.2	0.7–1.3
Not applicable[n]	Magnetite concrete	86	7–13	0.063–0.212	11.3–128	0.26–17.6	1.7–3.3	0.8–2.6	0.4–0.9
Not applicable[o,p]	Cement mortar	3	12	0.075	128	0.10–0.17	3.2	2.8–3.0	0.8–0.9

[a] Sanchidrián et al. 2006; [b] Bond & Whitney 1959; [c] McKenzie 2003; [d] Olsson & Bergqvist 2002; [e] Gynnemo 1997; [f] Cunningham 2003; [g] Otterness et al. 1991; [h] Grimshaw 1958; [i] Moser et al. 2003a; [j] Moser et al. 2003b; [k] Moser et al. 2003c; [l] Kristiansen 1994; [m] Kristiansen 1995a; [n] Rustan & Naartijärvi 1983; [o] Svahn 2002; [p] Svahn 2003.
[1] Limestone qualities in Klinthagen. Stroma: stromatoporoid; crin: crinoidal; mass: massive reef; frag: fragmented.

Table 1b. Fragmentation data: Crushed and milled material.

Mine or quarry site, or rock origin	Rock, equipment	No. of sets	n_p	x_{min} (mm)	x'_{max} (mm)	p_{min} (%)	r_x	r_p	s_L
3 Primary crusher (42 data sets)									
El Alto, Spain[a]	Limestone, toothed roller crusher	1	17	0.063	63	1.14	3.0	1.9	0.6
Kållered, Sweden[b]	Gneiss	10	14	0.074	40	4.9–7.4	2.7	1.1–1.3	0.4–0.5
Tampomas, Indonesia[c]	Andesite, jaw crusher	2	20–21	0.075	200–300	0.47–1.06	3.4–3.6	2.0–2.3	0.6–0.6
Klinthagen, Sweden[d]	Limestone, toothed roller crusher	28	9–18	0.063–8	200–300	0.13–18.8	1.4–3.7	0.7–2.9	0.5–0.8
Not reported[e]	Crusher	1	10–10	1.18	31.5	11.2	1.4	1.0	0.7
4 Secondary and tertiary crusher (50 data sets)									
Norway[f,g]	Anorthosite, gyratory crusher	4	12	0.075	32	0.60–0.81	2.6	1.6–1.7	0.6
Tampomas, Indonesia[c]	Andesite, gyratory crusher	15	16–20	0.075	63–200	0.20–1.6	2.9–3.4	1.8–2.7	0.6–0.9
Tampomas, Indonesia[c]	Andesite[1]	31	8–18	0.075–9.5	28–100	0.04–4.4	1.0–3.1	1.4–3.4	0.5–2.9
5 Mill (43 data sets)									
Tampomas, Indonesia[c]	Andesite, rod mill	9	9–12	0.075	6.3–19	1.05–6.2	1.9–2.4	1.2–2.0	0.6–0.8
Not reported[e]	AG mill	1	12	0.038	1.4	23.3	1.6	0.6	0.4
Not reported[e]	Hydrocyclone overflow	1	12	0.009	0.3	19.5	1.5	0.7	0.5
McCoy mine, USA[h,i,j]	Limestone, single particle roll mill	16	12	0.074	3.36	4.1–12.7	1.7	0.9–1.4	0.5–0.8
McCoy mine, USA[h,i]	Limestone, ball mill	16	11–12	0.074	2.38–3.36	8.3–34.6	1.5–1.7	0.5–1.1	0.3–0.7
6 Crusher and mill (OCS[2]) (38 data sets)									
Eibenstein, Austria[k]	Amphibolite	13	5–13	0.04	1–40	0.57–18.6	1.4–3.0	0.7–2.2	0.5–0.8
El Alto, Spain[k]	Limestone	5	5–13	0.04	1–40	0.46–11.7	1.4–3.0	0.9–2.3	0.7–0.8
Klinthagen, Sweden[k]	Limestone-stroma[3]	5	5–13	0.04	1–40	0.98–14.0	1.4–3.0	0.9–2.0	0.6–0.7
Klinthagen, Sweden[k]	Limestone-crin[3]	5	5–13	0.04	1–40	0.91–16.9	1.4–3.0	0.8–2.0	0.5–0.7
Klinthagen, Sweden[k]	Limestone-mass[3]	5	5–13	0.04	1–40	0.94–14.6	1.4–3.0	0.8–2.0	0.6–0.7
Klinthagen, Sweden[a]	Limestone-frag[3]	5	5–13	0.04	1–40	2.0–16.2	1.4–3.0	0.8–1.7	0.5–0.6

[a] Sanchidrián et al. 2006; [b] Gynnemo 1997; [c] Hänninger et al. 1988; [d] Moser 2003; [e] Napier-Munn et al. 1996; [f] Kristiansen 1995a; [g] Kristiansen 1995b; [h] Chi 1994; [i] Chi et al. 1996; [j] Fuerstenau et al. 1997; [k] MU Leoben 2002.
[1] Different materials sampled in the crushing plant have been included in this group: gyratory crusher product mixed with cone crusher recycled product, impact crusher feed (cone product partially sieved off) and product, and product partially sieved off (feed to a rod mill).
[2] Steiner's (1991) "optimized comminution sequences", consisting of several stages of crushing and milling with a small scale reduction plus classification and feedback.
[3] Limestone qualities in Klinthagen. Stroma: stromatoporoid; crin: crinoidal; mass: massive reef; frag: fragmented.

fitting code has been programmed in Matlab (2011) using non-linear function minimization routines. Two types of algorithms have been used: Levenberg-Marquardt (L-M) with a trust-region reflective method (Coleman & Li 1994, 1996) and Nelder-Mead simplex search method—a direct search method (Lagarias et al. 1998) that does not use numerical or analytical gradients—. The minimization problem is prone to reaching local minima, regardless of the algorithm used to solve it, so that the final minimum of the sum of squares, and the solution point, is very much dependent on the initial guess, this being especially true for functions of more than two parameters. In order to overcome this, each minimization problem (i.e. a data set and a distribution function) has been run repeatedly with different initial points, randomly generated within feasible intervals of the variables. The algorithm is assumed to have found the global minimum when repeated minimizations (from different initial guess vectors) reach the same solution and the same sum of squares within a given tolerance. The maximum number of minimizations was $1000 \cdot n_v$, with n_v being the number of parameters of the function being fitted.

Figure 1 gives an overview of the number of minimizations required for the various distributions and minimization algorithms, by means of boxplots of the ratio of the number of minimizations to the number of parameters of the distribution; circles show the 95% percentiles. WRR seems especially easy to fit with both algorithms tested; similar behavior is observed in LGN with L-M and LGL with direct search; TLGL appears to be the easiest-to-fit three-parameter function when using L-M, but one of the most difficult with direct search. GIL and TGIL are hard to converge to a global minimum, the three-parameter version not being worse than the truncated, four-parameter one. All truncated distributions are faster fitted with L-M than direct search, except TGIL and SWE. GRA and LGL are faster fitted with direct search and LGN with L-M.

For most of the distributions, the performance of both optimization techniques is very similar in

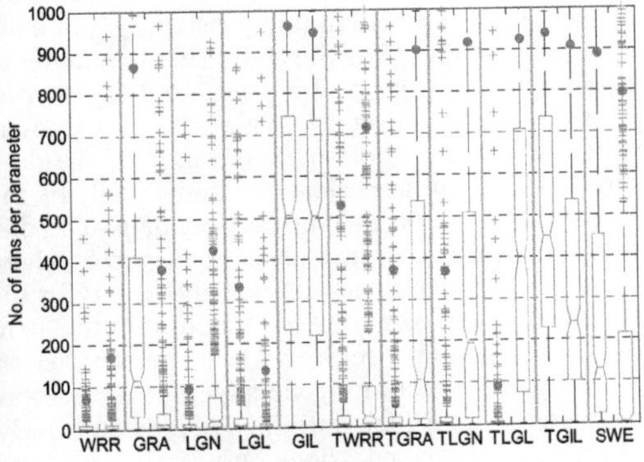

Figure 1. Number of runs per parameter. For each distribution function, left box: Levenberg-Marquardt; right box: simplex search.

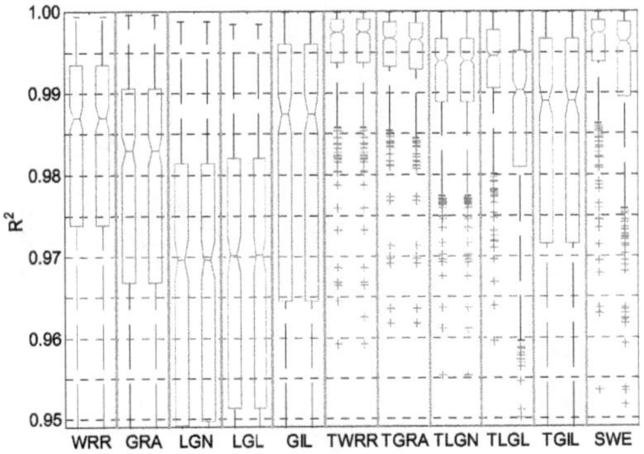

Figure 2. Determination coefficient. For each distribution function, left box: Levenberg-Marquardt; right box: simplex search.

obtaining the global (or, at least, the "best") minimum of the squares sum; this is analyzed in Figure 2 where the determination coefficient of the fits, R^2, is plotted. The cases where there is a noticeable difference are the truncated log-logistic and the Swebrec, in both cases in favor of the L-M method. In order to ensure that the objective function (the sum of squares) reaches the best possible minimum, the distributions have been fitted with both methods, and the best solutions have been retained.

4 RESULTS AND DISCUSSION

The determination coefficients in Figure 2 give a first indication of the global quality of the fits, the truncated functions outdoing their infinite versions as was to be expected since the former encompass one more parameter than the latter. In principle, it appears that the truncated WRR and GRA, together with the SWE, get the higher determination coefficients. The median values (using only the higher R^2 for each data set with any method) are given in Table 2; they are 0.9975 for the TWRR and 0.9974 for the SWE, which are statistically the same to a 0.01 significance; TGRA's R^2 is 0.9968, close to the two above but already statistically different. Besides the determination coefficients, Table 2 gives the $1-R^2$ values, which represent the fraction of variance unexplained by the fits. Truncation reduces this variance by a factor (given in the rightmost column of Table 2) of more than five in all cases except the Gilvarry, for which the truncation appears not to be really effective.

A convenient form of testing the behavior of the different functions across the whole range of the data is to calculate the differences between the actual sizes and the ones calculated from the fits at a given passing; logarithmic differences are used for this purpose:

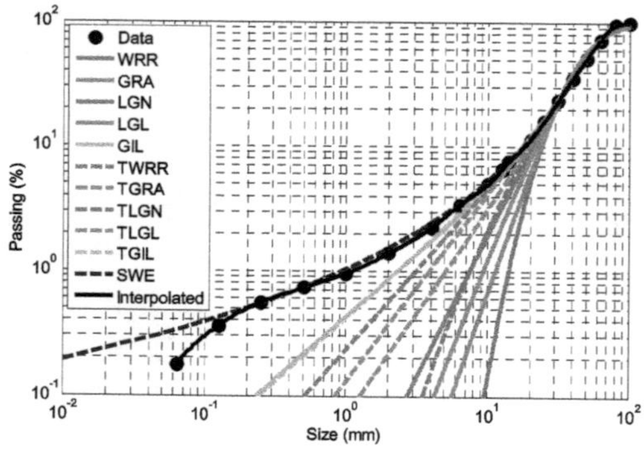

Figure 3. Example of fits.

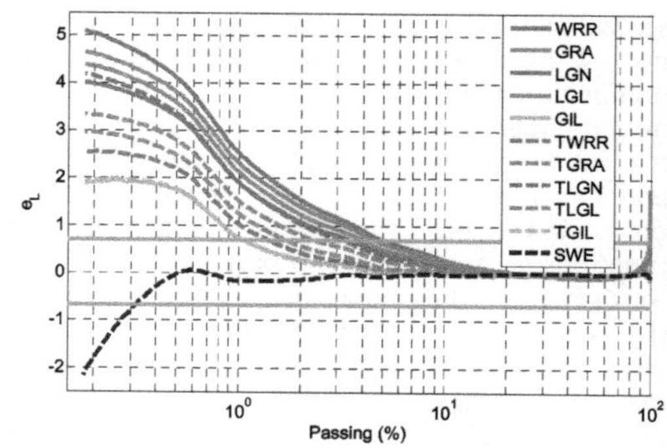

Figure 4. Log errors in size of the fits shown in Figure 3. Horizontal lines are ±ln 2 (relative errors 100 or −50%).

Table 2. Determination coefficients. Medians.

	R^2		$1-R^2$		
	Truncated	Plain	Truncated	Plain	Ratio*
WRR	0.9975	0.9870	0.0025	0.0130	5.279
SWE	0.9974	–	0.0026	–	–
GRA	0.9968	0.9829	0.0032	0.0171	5.274
LGL	0.9946	0.9703	0.0054	0.0297	5.519
LGN	0.9940	0.9697	0.0060	0.0303	5.031
GIL	0.9890	0.9875	0.0110	0.0125	1.140

* Ratio of unexplained variances of the plain (infinite variable) function to the truncated one.

$$e_L = \ln(x_p^*/x_p) \tag{10}$$

where x_p is the size at a percentage passing p in a given data set and x_p^* the size obtained from a given distribution function fitted to it:

$$x_p^* = F_{CDF}^{-1}(p) \tag{11}$$

The logarithmic error can be transformed into relative error as follows:

$$e_r = (x_p^* - x_p)/x_p = x_p^*/x_p - 1 = \exp(e_L) - 1 \tag{12}$$

or:

$$eL = \ln(e_r + 1) \tag{13}$$

To facilitate the analysis of errors, the method used by Sanchidrián et al. (2009a, b, 2012) has been followed: the original data are interpolated by Hermite polynomials (cubic polynomials matching the extremes of each data interval and with continuous derivative in those) in log-log space with the support points in the p axis spaced in such a way that the distribution of these replicates the distribution of passing points from all data sets; the x_p values are the interpolates for each support p point. This procedure allows the set of passing values used for the error calculations to be the same for all data sets (within the passing range of each one). Figure 3 shows, as a matter of illustration, the fits for one of the lab-scale amphibolite blasts; the interpolated curve is also shown (dark solid line running through all data points). Figure 4 shows the log errors for the fits in Figure 3. Two thousand passing values have been used for the support set so that the errors have a continuous-like appearance when plotted against the passing.

The different behavior of the fits across the range is apparent. This has been studied following the methodology used by Sanchidrián et al. (2012) by calculating the root mean squared logarithmic errors for the four zones mentioned in section 1:

$$RMSe_L = \left\{ \frac{1}{(p_s - p_i)} \int_{p_i}^{p_s} e_L^2 \, dp \right\}^{1/2} \tag{14}$$

where p_i and p_s are respectively the lower and upper limits of each passing zone (the p_i value for the very fines is the minimum passing, p_{min}, of each data set, and the p_s for the coarse is the maximum passing, p_{max}). If p_{min} and p_{max} are used together as p_i and p_s in Equation 14, a global mean error is obtained for the whole passing range of the data set. The integrals are calculated numerically by the trapezoidal rule on the support set of p. The distributions of $RMSe_L$ are given in Figure 5, where errors for the four zones defined above and for the full range are shown for each function in the form of box plots; solid circles are the 95 percentiles.

Truncated functions generally provide a more accurate fit than their infinite versions; curiously, Gilvarry's truncated does not add up any significant advantage to its three-parameter sister. This behavior may be connected with the fact that the three parameters of this function are all scale

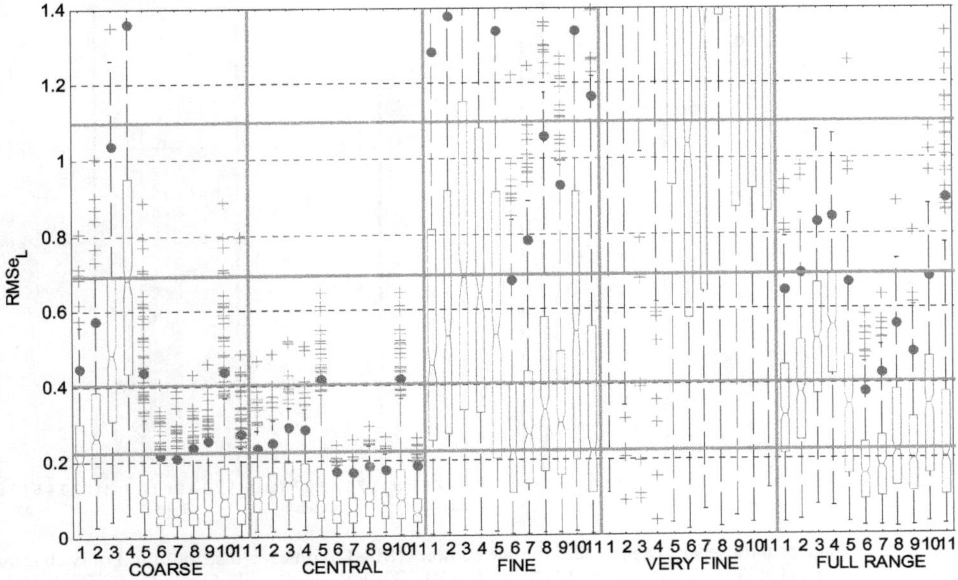

Figure 5. Distributions of root mean squared errors for all data sets at different passing zones. Solid circles are the 95 percentiles. For each zone, the order of the functions (1–11) is: WRR, GRA, LGN, LGL, GIL, TWRR, TGRA, TLGN, TLGL, TGIL and SWE. Horizontal lines are ln 1.25, ln 1.50, ln 2 and ln 3, which correspond to 25, 50, 100 and 200% relative errors respectively.

factors, so that adding one more (the maximum size is actually another scale factor) does not help to make the function more flexible, and an adjustable shape factor is missing (there are three shape factors in this function: 1, 2 and 3, but they are all fixed). The plain Gilvarry itself has a worse performance than any of the three-parameter functions compared. Size errors for the truncated functions (except Gilvarry) are generally (according to the 95% percentiles) less than about 25% in the coarse and central ranges, but they rise in the fines up to 100 to 200%; median errors are well below 10% in the coarse and central, and between 25 and 50% in the fines. In the very fine range, errors are usually wild. According to this result, the functions can be ranked in order of increasing median of the $RMSe_L$ in each range (functions mentioned are those for which the medians cannot be said to be different, from the Kruskal-Wallis test at 0.01 significance):

- Coarse: TGRA, TWRR, TLGN, SWE, TLGL. Median $RMSe_L$ 0.06–0.07 (relative error 6 to 7%).
- Central: TWRR, SWE, TGRA. Median $RMSe_L$ 0.06–0.07 (relative error 6 to 7%).
- Fine: TWRR, SWE, TGRA. Median $RMSe_L$ 0.22–0.27 (relative error 25 to 30%).
- Very fine: TWRR. Median $RMSe_L$ 1.04 (relative error 180%).
- Full range: TWRR, TGRA. Median $RMSe_L$ 0.16–0.18 (relative error 17–20%).

In order to investigate any evidence of a distinct behavior of the functions with fragmented materials of different origin, the number of data sets for which each function is the one with a lower $RMSe_L$ has been counted for each of the six groups of data (see Tables 1a–b) and each passing zone. The result is plotted in Figure 6, which shows the proportion (in percentage) of the data sets of a group for which a distribution gives the lower $RMSe_L$ for each passing zone. The global qualifications (all data sets considered) are also shown in the rightmost plot. Table 3 summarizes, for each data group and passing zone, the distributions with higher scores in Figure 6. Functions quoted are those for which the null hypothesis that their proportions, treated as multinomial variables, are equal to the higher one is not rejected at 0.05 significance.

Truncated Weibull and Swebrec compete as the best scoring functions of all tested here in virtually all materials and passing ranges without any significant preference for a particular material. Worth mentioning is the good behavior of the truncated log-logistic in the coarse range, in blasted (mine and lab) material, the truncated lognormal in the primary crusher material, coarse range, and the truncated Grady in the OCS material, coarse range. As full-range fitting functions, TWRR and SWE are preferred for all types of materials.

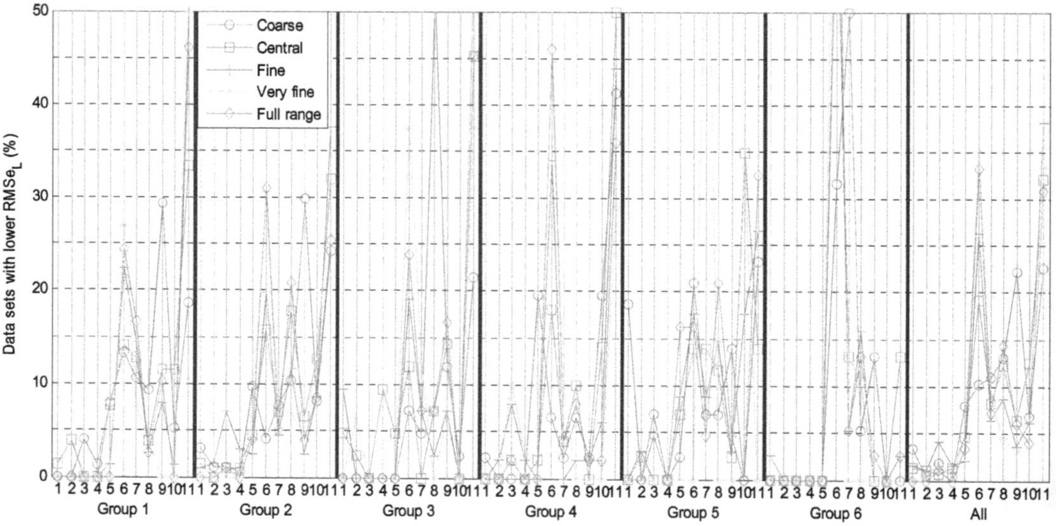

Figure 6. Percentage of data sets of each group that have lower RMSeL with each distribution. For each group, the order of the functions (1–11) is: WRR, GRA, LGN, LGL, GIL, TWRR, TGRA, TLGN, TLGL, TGIL and SWE. Results for the whole data sets are also shown.

Table 3. Performance for different type of data and passing zone. Figures in parentheses are the number of data sets for which the function has the lowest $RMSe_L$ in a given zone.

Group	Coarse	Central	Fine	Very fine	Full range
1 Blasted, mine	TLGL (22) SWE (14)	SWE (26) *TWRR (11)*	SWE (38) *TWRR (17)*	SWE (16) *TWRR (7)*	SWE (36) *TWRR (19)*
No. of sets	75	78	76	26	78
2 Blasted, lab	TLGL (58) SWE (47)	SWE (63) *TLGN (35)*	SWE (74) *TWRR (39)*	SWE (41) TWRR (40)	TWRR (61) SWE (50)
No. of sets	194	197	197	131	197
3 Primary crusher	TLGN (22) *SWE (9)*	SWE (19) *TLGL (6)*	SWE (23) *TWRR (8)*	TWRR (6) SWE (6)	SWE (19) TWRR (10)
No. of sets	42	42	42	16	42
4 Secondary and tertiary crusher	SWE (19) *GIL (9)* *TGIL (9)*	SWE (25) *TWRR (9)*	SWE (22) TWRR (17)	TWRR (17) SWE (11)	TWRR (23) SWE (18)
No. of sets	46	50	50	38	50
5 Mill	SWE (10) TWRR (9) WRR (8) TLGL (6)	TGIL (15) *TWRR (6)* *TGRA (6)* *SWE (6)*	SWE (9) TWRR (6) TGIL (6) TLGN (5)	TGRA (2) TGIL (1)	SWE (14) TLGN (9) GIL (7) TWRR (7)
No. of sets	43	43	34	3	43
6 Crusher and mill (OCS)	TGRA (19) TWRR (12)	TWRR (24) *TGRA (5)* *SWE (5)*	TWRR (28) *TLGN (6)*	TWRR (8)	TWRR (29) *TLGN (5)*
No. of sets	38	38	38	8	38
All	SWE (99) TLGL (97)	SWE (144) *TWRR (86)*	SWE (167) *TWRR (115)*	TWRR (78) SWE (74)	TWRR (149) SWE (138)
No. of sets	438	448	437	222	448

5 CONCLUSIONS

Eleven distributions have been compared in their capacity to fit rock fragment size distributions: Weibull-Rosin-Rammler, Grady, lognormal, log-logistic, Gilvarry, their truncated versions and Swebrec. Fits have been carried out to 448 data sets of fragmented rock of a variety of origins. Solutions are in general dependent on the initial point; a technique of multiple minimizations with random initial point between suitable bounds has been found appropriate to reach the global minimum in less than 1000 trials per unknown parameter. Two minimization algorithms have been applied, Levenberg-Marquardt and simplex direct search, with no general preference for one or the other observed. The minimum squared error solution from both algorithms has been, for each data set, selected.

Truncated functions are an excellent means of improving the determination coefficient, going in most cases from 0.97–0.99 for the ordinary distributions to 0.994–0.998 for the truncated ones; this increase corresponds to a five-fold reduction of the unexplained variance. The higher median determination coefficients are 0.9975 and 0.9974 for the truncated Weibull-Rosin-Rammler and Swebrec (itself a truncated distribution), respectively.

The quality of the fits across the passing range has been assessed in terms of differences of sizes calculated to the data values for a given passing, expressed in the form of logarithmic error, and this has been summarized for four arbitrary zones: coarse, central, fine and very fine. All truncated distributions, except Gilvarry, show a similarly good behavior in the coarse region (80–100%), with median relative size errors of 6 to 7%. Similar errors are found for the best distributions in the central range (20–80%), TWRR, SWE and TGRA. These three distributions are also the best ones in the fines region (2–20%), but relative errors are then 25 to 30%. In the very fines, errors are wild, with median 180% for the best distribution, the TWRR. As for the full range fitting, the TWRR and TGRA obtain a median error of 17–20%.

There is no significant distinction of the distributions when fitting data from different materials. The TWRR and SWE are usually the best fitting in most of the materials and ranges. In the coarse range TLGL does a good job for blasted material, TLGN for primary crusher one and TGRA for the OCS material.

ACKNOWLEDGEMENTS

We would like to thank Claude Cunningham and Cameron McKenzie for kindly sharing unpublished fragmentation data.

REFERENCES

Bond, F.C. & Whitney, B.B. 1959. The work index in blasting. In *Proc. 3rd US Symp. Rock Mech., Golden, CO. Quarterly Colorado School of Mines* 54(3): 77–82.

Chi, G. 1994. A study of comminution efficiency in relation to controlled blasting in mining. PhD thesis. Reno: University of Nevada.

Chi, G., Fuerstenau, M.C., Bradt, R.C. & Ghosh, A. 1996. Improved comminution efficiency through controlled blasting during mining. *Int. J. Mineral Proc.* 47: 93–101.

Coleman, T.F. & Li, Y. 1994. On the convergence of reflective Newton methods for large-scale nonlinear minimization subject to bounds. *Mathematical Programming* 67(2): 189–224.

Coleman, T.F. &, Li, Y. 1996. An interior, trust region approach for nonlinear minimization subject to bounds. *SIAM Journal on Optimization* 6(2): 418–445.

Cunningham, C.V.B. 2003. Personal communication.

Fuerstenau, M.C., Chi, G., Bradt, R.C. & Ghosh, A. 1997. Increased ore grindability and plant throughput with controlled blasting. *Mining Engineering* 49(12): 70–75.

Gilvarry, J.J. 1961. Fracture of brittle solids. I. Distribution function for fragment size in single fracture (Theoretical). *J. Appl. Phys.* 32(3): 391–399.

Gilvarry, J.J. & Bergstrom, B.H. 1961. Fracture of brittle solids. II. Distribution function for fragment size in single fracture (Experimental). *J. Appl. Phys.* 32(3): 400–410.

Grady, D.E. 1990. Particle size statistics in dynamic fragmentation. *J. Appl. Phys.* 68(12): 6099–6105.

Grady, D.E. & Kipp, M.E. 1985. Geometric statistics and dynamic fragmentation. *J. Appl. Phys.* 58(3): 1210–1222.

Grimshaw, H.C. 1958. The fragmentation produced by explosive detonated in stone blocks. In W.H. Walton (ed.), *Mechanical properties of non-metallic brittle materials, Proc. Conf. on Non-metallic Brittle Materials, London, April 1958*. London: Butterworth, pp. 380–388.

Gupta, A. & Yan, D.S. 2006. *Mineral processing design and operations*. Amsterdam: Elsevier.

Gynnemo, M. 1997. Investigation of governing factors in bench blasting. Full-scale tests at Kållered and Billingsryd. Publ. A84. Gothenburg, Sweden: Chalmers University, Department of Geology. In Swedish.

Hänninger, A., Larsson, L. & Slokenbergs, M. 1988. Investigation of results and capacities when crushing andesite. Dipl. Thesis no. 308, Hydraulics Lab. Stockholm: Royal Institute of Technology. In Swedish.

King, R.P. 2001. Modeling and simulation of mineral processing systems. Oxford: Butterworth-Heinemann.

Kristiansen, J. 1994. Blastability of rock, small-scale tests in rock blocks. Rep. 548095-4. Oslo: Norwegian Geotech. Inst. In Norwegian.

Kristiansen, J. 1995a. Blastability of rock, full-scale blasting tests with different hole diameters. Rep. 548095-5. Oslo: Norwegian Geotech. Inst. In Norwegian.

Kristiansen, J. 1995b. A study of how the velocity of detonation affects fragmentation and the quality of

fragments in a muckpile. *Proc. Explo'95, Publication series no. 6/95*. Carlton, Vic, Australia: AusIMM, pp. 437–444.
Lagarias, J.C., Reeds, J.A., Wright, M.H. & Wright, P.E. 1998. Convergence properties of the Nelder-Mead Simplex method in low dimensions. *SIAM Journal on Optimization* 9(1): 112–147.
Matlab 7.13. 2011. Natick, MA: The MathWorks Inc.
McKenzie, C. Personal communication. 2003.
Moser, P. 2003. Less fines production in aggregate and industrial minerals industry. In R. Holmberg (ed.), *Proc. 2nd World Conf. on Explosives and Blasting, Prague, 10–12 September 2003*. Rotterdam: Balkema, pp. 335–343.
Moser, P., Grasedieck, A., Arsic, V. & Reichholf, G. 2003a. Charakteristik der Korngrössenverteilung von Sprenghauwerk im Feinbereich. *Berg- und Hüttenmännische Monatshefte* 148:205–216. In German.
Moser, P., Grasedieck, A., du Mouza, J. & Hamdi, E. 2003b. Breakage energy in rock blasting. In R. Holmberg (ed.), *Proc. 2nd World Conf. on Explosives and Blasting, Prague, 10–12 September 2003*. Rotterdam: Balkema, pp. 323–334.
Moser, P., Grasedieck, A., Olsson, M. & Ouchterlony, F. 2003c. Comparison of the blast fragmentation from lab-scale and full-scale tests at Bårarp. In R. Holmberg (ed.), *Proc. 2nd World Conf. on Explosives and Blasting, Prague, 10–12 September 2003*. Rotterdam: Balkema, pp. 449–458.
MU Leoben. 2002. The natural breakage characteristics (NBC) and energy register functions for the rock mass of the quarries Nordkalk, Cementos Portland and Hengl Bitustein. Less fines, EU Project GRD-2000–25224, Tech. Rep. 33.
Napier-Munn, T.J., Morell, S., Morrison, R.D. & Kojovic, T. 1996. Mineral comminution circuits – Their operation and optimisation. In *JKMRC Monograph Series in Mining and Mineral Processing*. Indooroopilly, Queensland, Australia.
Olsson, M. & Bergqvist, I. 2002. Fragmentation in quarries. In *Proc. Disc. Meeting BK 2002, Swedish Rock Construction Comm., Stockholm, Sweden*, pp. 33–38. In Swedish.
Otterness, R.E., Stagg, M.S., Rholl, S.A. & Smith, N.S. 1991. Correlation of shot design parameters to fragmentation. In *Proc. 7th Ann. Symp. on Explosives and Blasting Research, Las Vegas, Nevada, USA, 6–7 February 1991*, pp. 179–190.
Ouchterlony, F. 2005a. The Swebrec© function, linking fragmentation by blasting and crushing. *Mining Technology (Trans Inst Min Metal A)* 114: A29–44.
Ouchterlony, F. 2005b. What does the fragment size distribution of blasted rock look like? In R. Holmberg (ed.), *Proc. 3rd World Conf. on Explosives and Blasting, Brighton, 13–16 September*, pp. 189–199.
Rosin, P. & Rammler, E. 1933. The laws governing the fineness of powdered coal. *J. Inst. Fuel* 7: 29–36.
Rustan, A. & Naarttijärvi, T. 1983. The influence from specific charge and geometric scale on fragmentation. Rep. FG8322. Luleå, Sweden: Swedish Mining Research Foundation.
Sanchidrián, J.A., Segarra, P., López, L.M. 2006. A practical procedure for the measurement of fragmentation by blasting by image analysis. *Rock Mech. Rock Eng.* 39(4): 359–382.
Sanchidrián, J.A., Segarra, P., Ouchterlony, F. & López, L.M. 2009a. On the accuracy of fragment size measurement by image analysis in combination with some distribution functions. *Rock Mech. Rock Eng.* 42(1): 95–116.
Sanchidrián, J.A., Ouchterlony, F., Segarra, P., Moser, P. & López, L.M. 2009b. Evaluation of some distribution functions for describing rock fragmentation data. In J.A. Sanchidrián (ed.), *Proc. 9th Int. Symp. on Rock Fragmentation by Blasting (Fragblast 9), Granada, Spain, 13–17 September 2009*. Leiden: CRC Press/Balkema, pp. 239–248.
Sanchidrián, J.A., Ouchterlony, F., Moser, P., Segarra, P. & López, L. 2012. Performance of some distributions to describe rock fragmentation data. *Int. J. Rock Mech. & Min. Sci.* 53: 18–31.
Sil'vestrov, V.V. 2004a. Fragmentation of a steel sphere under a high-velocity impact on a highly porous thin bumper. *Combust. Expl. Shock Waves* 40(2): 238–52.
Sil'vestrov, V.V. 2004b. Application of the Gilvarry distribution to the statistical description of fragmentation of solids under dynamic loading. *Combust. Expl. Shock Waves* 40(2): 225–237.
Steiner, H.J. 1991. The significance of the Rittinger equation in present-day comminution technology. In *Proc. Int. Min. Proc. Congr., Dresden, Germany, Vol. I*, pp. 177–188.
Svahn, V. 2002. Generation of fines around a borehole: a laboratory study. In *Proc. 7th Int. Symp. on Rock Fragmentation by Blasting, Beijing, 11–15 August 2002*. Beijing: Metallurgical Industry Press, pp. 122–127.
Svahn, V. 2003. Generation of fines in bench blasting. Licenciate thesis, Pub. A104. Gothenburg, Sweden: Chalmers University, Department of Geology.
Weibull, W. 1939. A statistical theory of the strength of materials. *Ingeniörvetenskapsakademiens Handlingar* 151: 1–45.
Weibull, W. 1951. A statistical distribution function of wide applicability. *J. Appl. Mech. ASME* 18: 293–297.

Applications of fragmentation measurement

Measurement and Analysis of Blast Fragmentation – Sanchidrián & Singh (Eds)
© 2013 Taylor & Francis Group, London, ISBN 978-0-415-62140-3

Assessment of objective based blast performance: Ranking system

V.R. Sastry & K.R. Chandar
Department of Mining Engineering, National Institute of Technology, Karnataka, Surathkal, Mangalore, India

ABSTRACT: Rock fragmentation using chemical energy is commonly followed for fragmenting and displacing the rock mass in mining, civil and other excavation projects. In case of mines, the objective of rock fragmentation is to get a uniform sized material to suit the subsequent operations like loading, hauling and crushing. In case of civil engineering projects like canals, tunnels, it is to provide the space for some facilities and the material produced may not be having a specific use. In case of blasting near to the human habitats or important structures like dams, ground vibrations generated should be within permissible safe limits, along with proper fragmentation. Cost reduction is also a prime factor in all the cases. In general, assessment of blast performance refers to fragmentation assessment. However, based on the objective, the approach to blast design may be different. An attempt is made through this paper to discuss different blast performance assessment methods, through a case study and a ranking system is proposed. Field studies were carried out in a limestone mine in southern India, using eight controlled blasts by changing the initiation system, initiation pattern and maintaining same blast geometry parameters. Blast performance was assessed based on fragmentation using digital image processing technique, energy distribution, and productive yield and powder factors resulting from the blasts. Rankings were given to different blasts as per the performance.

1 INTRODUCTION

Explosive is the most commonly used energy to fragment the rock mass in mining and civil engineering projects. Main objective of blasting in mining projects is to obtain the maximum yield with desired fragmentation in a safer manner with minimum side effects like ground vibrations, noise and fly rock, while in civil engineering projects it is to create space. Assessment of each blast is necessary keeping the objective of the blast in view. Blast results can be categorized into two groups of desirable fragmentation and unwanted results like ground vibrations, noise and fly rock. In addition, there are some more minor undesirable results like back break, toe formation etc. A number of factors influence the blast results, which can be grouped into controllable and uncontrollable factors. Parameters like explosive type, burden, spacing, sub drilling, stemming, delay timing, charge weight per delay, initiation system, initiation pattern etc., may be grouped under controllable parameters. Geological parameters like mineralogy, lithology, structural discontinuities and physico-mechanical properties of rock mass come under uncontrollable factors, as they are given by nature. Initiation system is one of the major parameters that influence the blast results considerably. Assessment of blast results in terms of fragmentation forms a major basis while evaluating the influence of various parameters.

The term fragmentation refers to post blast size distribution of the broken rock mass. By considering the overall system performance, optimum fragmentation may, therefore, be defined as the fragmentation that results in the minimal overall costs for drilling, blasting, loading, transporting and crushing (Mackenzie 1966, Greenland & Knowles 1969, Nielsen 1985). Da Gama & Jimeno (1993) indicated that at optimum fragmentation, environmental impacts will also be minimum. Fragmentation obtained in the blasting process influences the downstream costs like loading, transportation, processing cost, etc. The fragment size should be suitable for further handling and processing equipment in order to reduce the total production cost. There is a significant evidence that blasting does affect crushing and grinding results (Eloranta 1995, Paley & Kojovic 2001). The use of greater energy input in the blasting unit operation will be cheaper than expanding the energy into downstream operations (Workman & Eloranta 2003). Improved fragmentation accomplished in blasting not only reduces the work load in crushing and grinding, but also improves loading rates and reduces maintenance in the mine (Eloranta 1995).

The concept of optimum fragmentation is critical for optimizing a drilling and blasting program that minimizes the entire cost for a mining operation (Hustrulid 1999, Kanchibotla 2001, La Rosa 2001). Mechanical crushing and grinding are particularly expensive than breaking the rock using explosives energy (Eloranta 1997, Simangunsong et al. 2003, Katsabanis et al. 2003). Process of crushing rock material can be five times that of blasting the rock material (Chitombo et al. 2001). Though the drilling and blasting cost increases with the degree of fragmentation, the cost of subsequent operations like loading, transportation and cursing cost will come down. In overall, the total product cost minimizes (Fig. 1).

Various methods area available to assess the fragmentation like screening, explosive consumption in secondary blasting, boulder counts, bridging delay at crusher, shovel and crusher energy consumption and image processing. Among all these, screening is the only direct and more accurate method. But, in the case of large projects it is cumbersome, time consuming and costlier. Image processing technique is a good alternative which can be used without disturbing the production process.

If rock size uniformity is high and thickness of layer is low, the image-processing program is useful and efficient. However, if the uniformity of rock size is low and thickness of layer is significant, the user should be especially careful in accepting the results of image analysis (Cunningham 1996).

It is very difficult and hard to obtain accurate estimates of rock fragmentation after blasting. Following are the main reasons for errors in using image analysis programs (Liu & Tran 1996):

- Image analysis can only process what can be seen with the eye.
- Image analysis programs cannot take into account the internal rock, so the sampling strategies should be carefully considered.
- Analyzed particle size can be over-divided or combined, i.e, larger particles can be divided into smaller particles and smaller particles can be grouped into larger particles. This is a common problem in all image-processing programs. Therefore, manual editing is required.

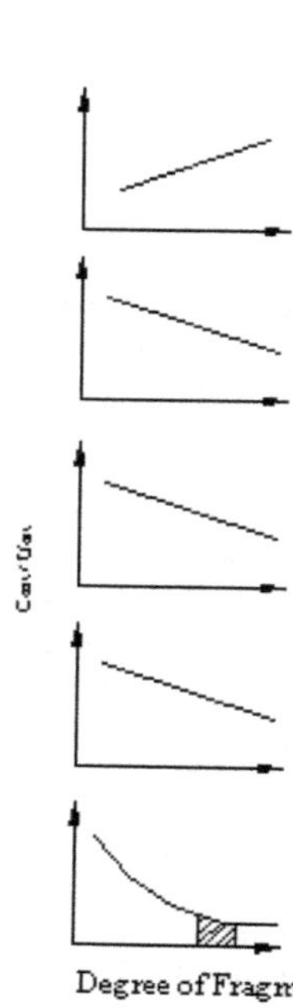

Figure 1. Effect of fragmentation on downstream operations (Mackenzie 1966).

In addition, the rock mass should not only be adequately fragmented but it should also be displaced into a suitable muck pile, providing better conditions for digging and hauling equipment (Harries 1987). In general, in a properly designed blast around a meager 15 per cent of the energy is doing useful work and the remaining is wasted in generating side detrimental effects (Hagan 1973). Berta (1990) approximately estimated the distribution of utilization of total explosive energy into following categories:

Fracture in-situ: <1%
Breakage: 15%
Displacement: 4%
Crushing in the vicinity of the hole: 1.5 to 2%
Fly-rock: <1%
Deformation of solid rock behind the shot: <1%
Ground vibrations: 40%
Air blast/noise: 38 to 39%

Various researchers have tried to correlate the fragmentation with productivity of loaders, efficiency of crushers and grinders etc. Grant & Dutton (1983) found no correlation between fragment size and loading rate until a boulder was encountered. Presence of boulders significantly increases the loading cycle time. Nielsen (1987) conducted model studies of loading equipment as a function of rock fragmentation and observed fairly good linear correlation. Chung et al. (1991) studied shovel digging time to obtain data on the effects of explosive energy consumption on the shovel productivity. McGill & Freadrich (1994) conducted loader productivity studies in limestone and sandstone mines. Loader cycle time was compared with the fragmentation of the muck pile in two different rock formations. They concluded that the cycle time is directly related to the fragmentation size. Sastry & Chandar (2004) conducted studies in limestone mines to study the influence of initiation system on shovel cycle time and they found that the shovel cycle time is around 10 per cent more for detonating cord initiated blasts compared to shocktube initiated blasts. Hanspal et al. (1995) reviewed the influence of muck pile characteristics on performance of loaders. Singh & Narendrula (2006) compared loader performance with different parameters including muck pile, and found that good displacement and looseness of the muck pile gives better loading conditions for the loaders.

2 EXPLOSIVE ENERGY

Explosive energy is rated in a number of ways, obtained either from theoretical calculations or from experimental tests. However, it is very difficult to determine the amount of explosive energy transferred to the rock and converted into efficient work in the application of rock blasting. Although the measurement of some of the effects of the explosive energy in rock is customary (vibration, fragmentation and, to some extent, rock movement), they are usually conducted for blast control purpose and the results are rarely cast in terms of their energy content. The reason for this could be that it is not the energy consumption that matters, but rather than the end effects, i.e., degree of fragmentation, throw and vibration levels. Data and estimations on energy components in rock blasting are thus limited to a few researchers. Berta (1990), Spathis (1999) and Ouchterlony et al. (2003) estimated the amount of energy transformed in kinetic energy of the rock, fracture generation and seismic waves.

Energy released by an explosive, can be grouped into gaseous energy and shock energy, working on the surrounding strata resulting in fragmenting the medium through various breakage mechanisms. Some of these mechanisms are responsible for (a) fracturing energy, that ultimately is responsible for creating new surfaces in the rock fragments, (b) energy transferred as shock waves into the rock mass propagating as seismic waves or ground vibrations beyond the fragmentation zone, and (c) energy to displace the fragmented material and form the muck pile, that appears as kinetic energy. This energy partitioning is related to the characteristics of explosives and the strata, and to some extent to the blast geometry. The energy balance of the blast can thus be expressed as (Spathis 1999):

$$EE = EF + ES + EK + ENM$$

where,
EE = Explosive energy
EF = Fragmentation energy
ES = Seismic energy
EK = Kinetic energy
ENM = Energy forms not measured

Ground vibrations and air blast resulting from rock blasting are the unwanted manifestation of explosive energy. Now a days blasting operations are taking place near to human habitats, and domestic and historic structures also. Higher ground vibration levels indicate greater wastage of explosive energy.

While achieving good yield with desirable fragmentation, it is also important to minimize the wastage of explosive energy manifesting in the form of ground vibrations in order to optimize the explosive energy. A systematic study was taken up to combine both the fragmentation and the energy carried by ground vibrations for assessing the performance of blasts. Blast results were analyzed in terms of fragmentation, energy distribution, productive yield and powder factor. A new method is proposed for effective assessment of blast performance based on energy distribution.

3 INVESTIGATIONS

Field studies were carried out in a limestone mine in southern India. The daily production of the mine is 3000 t. Deposit is of good quality with an average $CaCO_3$ (calcium carbonate) content of about 83 percent. Conventional benching method is adopted to excavate the limestone deposit, with two operating benches of 8 m height each. Blastholes are of 115 mm diameter. Burden to spacing pattern of 3.5 × 5.0 m is followed in this project. Cartridged slurries and ANFO are used as primer and column charges respectively. A view of the mine is shown in Figure 2.

In total, eight experimental blasts were studied using detonating cord and nonel based shocktube systems of initiation, with—single row and row-to-row pattern of initiation (Fig. 3). For each

Figure 2. View of the limestone mine.

Z	25	50	75
O	O	O	O

a. single row

Z	Z	Z	Z
O	O	O	O
25	25	25	25
O	O	O	O

b. two rows

Figure 3. Blast patterns.

Table 1. Details of blasts conducted.

Sl. no	B × S × H m × m × m	No.	Charge/hole kg	MCD kg	TC kg	D m	SD m/kg$^{0.5}$	PPV mm/s	Initiation system	Initiation pattern
1	3.5 × 5.0 × 8.0	04	40	40	160	60	9.49	16.2	Shocktube	Single row
						90	14.23	11.9		
2	3.5 × 5.0 × 8.0	04	40	40	160	60	9.49	14.8	Shocktube	Single row
						90	14.23	11.2		
3	3.5 × 5.0 × 8.0	04	40	40	160	60	9.49	22.1	D-cord	Single row
						90	14.23	15.4		
4	3.5 × 5.0 × 8.0	04	40	40	160	60	9.49	21.6	D-cord	Single row
						90	14.23	16.5		
5	3.5 × 5.0 × 8.0	08	40	160	320	60	4.74	24.1	Shocktube	Row-by-row
						90	7.12	18.6		
6	3.5 × 5.0 × 8.0	08	40	160	320	60	4.74	22.8	Shocktube	Row-by-row
						90	7.12	17.5		
7	3.5 × 5.0 × 8.0	08	40	160	320	60	4.74	43.4	D-cord	Row-by row
						90	7.12	37.5		
8	3.5 × 5.0 × 8.0	08	40	160	320	60	4.74	41.7	D-cord	Row-by row
						90	7.12	36.2		

B × S × H: Burden × Spacing × Height of the bench; No: Number of blastholes; MCD: Maximum charge per blasthole; TC: Total charge; D: Distance; SD: Scaled distance; PPV: Peak particle velocity.

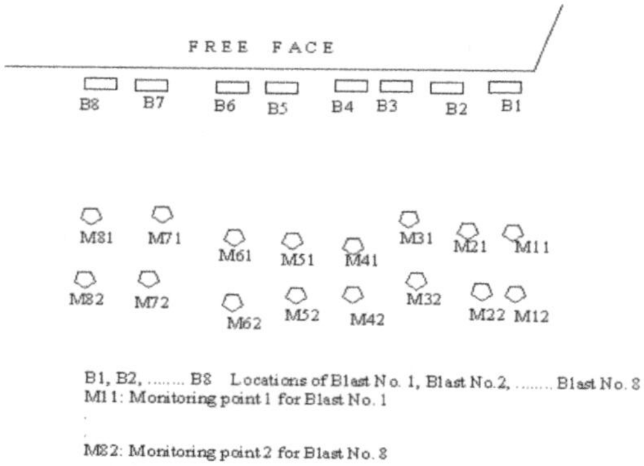

Figure 4. Location of blasts along with monitoring points.

initiation sequence, two blasts were carried out. Care was taken to conduct the blasts in the same bench almost side by side for repeatability of results.

Ground vibrations generated from blasting operations were monitored using two units of Minimate DS-077 of Instantel, Canada, at different distances from respective blast sites. Details of blasts monitored are given in Table 1. Details of blast locations along with monitoring points are shown in Figure 4. Performance of each blast was assessed for fragmentation, ground vibration, shock energy distribution and powder factor.

4 ANALYSIS OF BLAST PERFORMANCE

4.1 *Fragmentation analysis*

Fragmentation analysis was carried out using digital image processing technique. Muck piles were photographed using high shutter speed video camera by keeping known dimension calibrator on the muck pile. The images were processed using Fragalyst 2.0 software, which gives the output in the form of Rossin-Raamler distribution curve. Average fragment size (K_{50}) of muck pile was considered for comparison of results (Table 2). Average value of the repeated muck profiles was taken into consideration for analysis. Smaller K_{50} values indicate better fragmentation. Based on this criterion, Rank-1 was assigned to the blast resulting in lowest average fragment size. The next lower value was assigned Rank-2, and same procedure was repeated for all the blasts.

4.2 *Distribution of ground vibrations around the blastholes*

To assess the blast performance in terms of ground vibrations, the concept of propagation and distribution of ground vibration around the blastholes was taken up. JKSimBlast software was used for simulating all the blast patterns for analysis. Blast geometry parameters, explosive parameters and rock properties were input into the model. For analyzing the distribution of ground vibration, the area around the blast was categorized into zones of ground

Table 2. Average fragment size (K_{50}) of different blasts.

Initiation pattern	Average fragment size (K_{50})				No. of boulder per 1000 t			
	Shocktube		D-cord		Shocktube		D-cord	
	Individual	Average	Individual	Average	Individual	Average	Individual	Average
Single row	0.50		0.70		25	24	40	38
Single row	0.48	0.49	0.68	0.69	23		36	
Row-by-row	0.53		0.70		26	24	36	35
Row-by-row	0.50	0.52	0.70	0.71	22		34	

vibration intensity of 0 to 10 mm/s, 10 to 25 mm/s, 25 to 50 mm/s and greater than 50 mm/s. Area of each zone was estimated for comparison purpose. Areas of various zones with different intensities of ground vibration obtained are given in Table 3. Ground vibration distribution obtained for a typical blast is shown in Figure 5. Blast round resulting in lowest PPV value in the first zone of 0–10 mm/s was assigned Rank-1 and the Rank value increased as intensity of ground vibration increased.

Table 3. Ground vibration distribution zones for different blasts.

Pattern	Ground vibration Distribution (m^2) for different PPV ranges (mm/s)					
	D-Cord			Shocktube		
	>50	25–50	10–25	>50	25–50	10–25
Single row	390.5	481.7	1412.7	82.0	778.0	2889.5
Row by row	515.2	484.5	970.0	254.7	493.5	2395.2

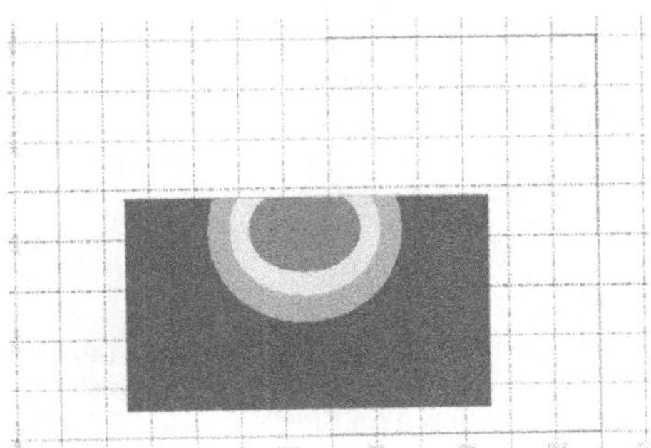

Figure 5. Distribution of ground vibrations around the blastholes.

4.3 *Energy distribution around the blastholes*

Once the explosive charge in blasthole is initiated, the energy dissipates into the rock mass as shock energy. The energy level is higher at the vicinity of blasthole and as distance from blasthole wall increases the energy available at any given point decreases. Rock mass gets fragmented when the energy available is greater than the breaking strength of rock. Based on this concept, the blast patterns were simulated using JKSimBlast software. Energy distribution around the blastholes was divided into four zones/ranges. The first one was carrying greater than 2 MJ/m^3 of energy, the second zone 1–2 MJ/m^3, third zone 0.25–1 MJ/m^3 and the last zone with less than 0.25 MJ/m^3 of energy. All these zones of energy resulting from the blasts were estimated and the values are given in Table 4. A typical output for a blast is shown in Figure 6. Rank-1 was assigned to the blast resulting in highest energy level. Shocktube initiation gave the highest energy at the vicinity of the blast site, with single row and row-by-row initiation pattern compared to detonating cord initiated blasts. Among shocktube initiated blasts, blast with two rows resulted in higher energy levels compared to a single row blast. Higher energy availability results in better fragmentation.

Table 4. Energy distribution around the blastholes.

Pattern	Area (m^2) for different energy ranges (MJ/m^3)					
	D-Cord			Shocktube		
	>2	1–2	0.25–1	>2	1–2	0.25–1
Single row	288.0	230.6	335.2	311.0	167.2	300.8
Row by row	318.0	370.6	472.0	381.0	182.2	426.0

4.4 *Productive yield*

Muck pile produced from the blast round should be suitable for excavators. A good muck pile profile

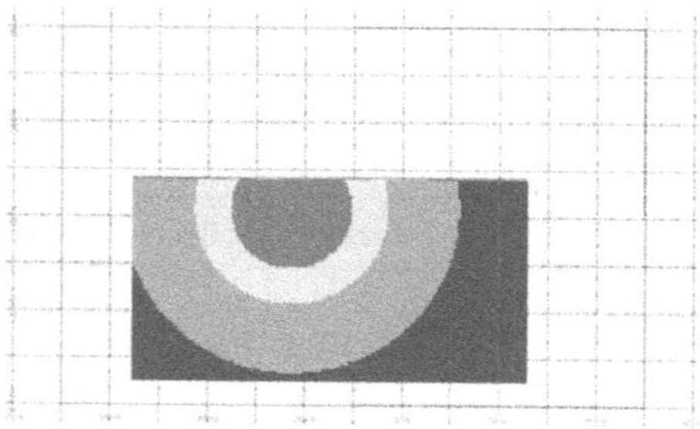

Figure 6. Energy distribution around the blastholes of a typical blast.

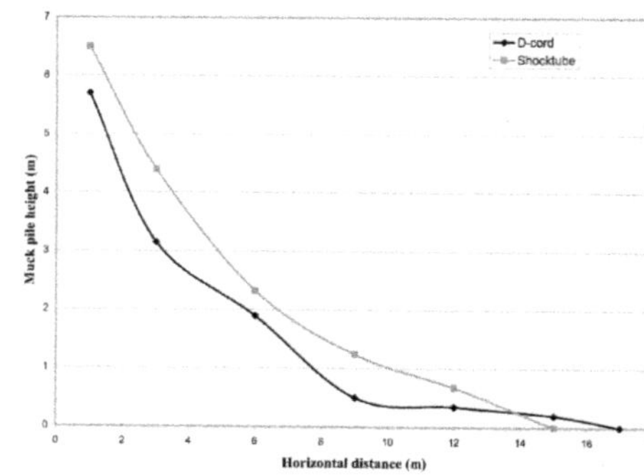

Figure 7. Muck pile profile of single row initiated blasts.

enhances the productivity of excavator. A typical blast profile is given in Figure 7. Total yield per blast was estimated based on muck pile survey and two new parameters were proposed namely, 'burden throw ratio' and 'muck pile height ratio'. Burden throw ratio is the ratio of muck pile throw distance to burden. Muck pile height ratio is the ratio of muck pile height to bench height. Both these ratios with a value of above 0.1 indicate desirable muck pile profile suitable to the excavators. The burden displacement ratio of different blasts is given Table 5. Muck pile height ratio is details are given in Table 6. Productive yield was estimated for all the blast rounds, considering the muck pile ratio and burden throw ratio of upto 0.1 (Table 7). The blast round giving the highest productive yield per blasthole was assigned Rank-1. Blast rounds with row by row initiation using shocktube system resulted in better productive yield.

Table 5. Percentage variation among muck pile throw distances of different blasts.

Initiation pattern	Throw distance (m)		Burden displacement ratio	
	Shocktube	D-cord	Shocktube	D-cord
Single row	15.4	16.2	4.40	4.62
Single row	15.0	19.0	4.28	5.42
Row to row	21.6	27.5	6.17	7.86
Row to row	21.5	28.1	6.14	8.03

Table 6. Muck pile height and muck pile ratio for row by row pattern of initiation.

Horizontal distance (m)	Muck pile height (m)		Muck pile height ratio	
	Shocktube	D-cord	Shocktube	D-cord
1	3.20	2.80	0.40	0.35
3	6.40	6.00	0.80	0.75
6	7.10	6.40	0.89	0.80
9	6.00	5.80	0.75	0.73
12	4.80	4.40	0.60	0.55
15	3.00	2.40	0.38	0.30
18	1.40	1.50	0.18	0.19
21	0.65	0.90	0.08	0.11
24	0.00	0.50	0.00	0.06
27	0.00	0.20	0.00	0.03

4.5 *Powder factor*

Powder factor is one of the important parameters used to assess the blast performance. Powder factor was calculated based on the total yield (Table 7) and total charge per blast (Table 1). Powder factor values in m^3/kg for each blast are given in Table 8. Blast with single row initiated by detonating cord gave the lowest powder factor. Within the pair of blasts, the variation in powder factors was very less.

A higher rank shows poor performance of the blast compared to a lower ranked one. Analysis showed that there is a good correlation among the parameters considered for analysis. Shocktube system of initiation resulted in better Average Fragment Size representing better fragmentation compared to detonating cord initiated blasts. Same trend was obtained with single and double row blasts. Similarly, the Productive Yield was better with shocktube initiated blasts, providing better muck pile profile, with lesser spread compared to detonating cord initiated blasts, which in turn results in better performance of the excavation equipment. The blasts initiated with shocktube system resulted in better energy distribution compared to detonating cord initiated blasts. Row by row initiation with detonating cord resulted in least productive yield and poorest energy distribution and fragmentation.

Table 7. Total yield and productive yield of the blasts.

Sl. no	No. of holes	Initiation system	Initiation pattern	Total yield (m³)	Productive yield (m³) Individual	Productive yield (m³) Average/blasthole
1	4	Shocktube	Single row	545	524	131.75
2	4	Shocktube	Single row	547	530	
3	4	D cord	Single row	555	496	123.25
4	4	D cord	Single row	550	490	
5	8	Shocktube	Row by row	1155	1052	132.25
6	8	Shocktube	Row by row	1150	1064	
7	8	D cord	Row by row	1175	995	123.87
8	8	D cord	Row by row	1172	987	

Table 8. Powder factor (m³/kg) for different blast patterns.

	D-cord		Shocktube	
Blast pattern	Individual	Average	Individual	Average
Single row	3.47	3.455	3.41	3.415
Single row	3.44		3.42	
Row by row	3.67	3.665	3.61	3.60
Row by row	3.66		3.59	

Table 9. Ranking of different blasts.

Initiation system/pattern	Based on K_{50} value	Based on PPV distribution	Based on energy distribution	Based on productive yield	Based on powder factor
Single row, shocktube	1	1	3	2	3
Single row, D-cord	3	3	4	4	4
Row by row, shocktube	2	2	1	1	2
Row by row, D-cord	4	4	2	3	1

5 OVERALL PERFORMANCE: RANKING SYSTEM

Among the different blast patterns with both shocktube and detonating cord initiation systems, to select a better one, quantifiable approach may be difficult as initiation system and pattern are not quantifiable parameters. So the results were assessed based on relative performance as per the rank allocated as shown in Table 9. Rank-1 shows best performance and Rank 4 (highest number) shows very poor performance and more unwanted results. Further, giving a cumulative ranking may not hold good as the objective of each blast may vary and assessing the blast performance is different, and each method carries a different weightage. Table 9 gives a guideline to select the best possible initiation system for the given geo-mining conditions.

6 CONCLUSIONS

Blasting operations are associated with desirable and some unwanted results. The unwanted effects may not be avoided completely, but can be reduced if the blast is designed effectively. Blast design should be carried out keeping the objectives of the blast in view. Subsequent assessment of the blast to verify whether it has fulfilled its objectives using a suitable assessment method is very important. In spite of having many conventional parameters, it becomes practically difficult to identify the performance of a blast accurately. An attempt is made through this paper to establish a concept based on parameters of fragmentation, ground vibration, energy distribution, productive yield and powder factor to assess performance of blasts. Concept of productive yield gives a good correlation with performance of the shovel. Energy distribution indicates the significance of ground vibrations resulting from a blast. Importance of each objective of the blast may vary from site to site. It is hoped that the approach discussed in this paper may become a guideline to select a suitable blast pattern and initiation system to achieve the required objectives.

REFERENCES

Berta, G. 1990. *Explosives: An engineering tool.* Milano: Italesplosivi.

Chitombo, G., Morrison, R. & Valey, W. 2001. Comminution optimization from mine to mill. *De Beers workshop on Communition and Mine to Mill*, pp. 71–79.

Chung, S.H., Lee, N.H. & Huter, C.J. 1991. Blast design analysis for optimizing productivity at INCO Ltd's Thompson Open Pit. *Proc. 17th Int. Conference on Explosives and Blasting Technique, Las Vegas, USA*, pp. 119–127.

Cunningham, C.V.B. 1996. Optimal fragmentation assessment—A technical challenge. *Proc. Measurement of Blast Fragmentation*. Rotterdam: Balkema.

Da Gama, C.D. & Jimeno, C.L. 1993. Rock fragmentation control for blasting cost minimization and environmental impact abatement. *Proc. 4th Int. Symp. on Rock Fragmentation by Blasting, Vienna*, pp. 273–280.

Eloranta, J. 1995. The selection of powder factor in large diameter blastholes. *Proc. 21st Annual Conference on Explosives and Blasting Research, Vol. I, Nashville, TN*, pp. 68–77.

Eloranta, J. 1997. The efficiency of blasting versus crushing and grinding. *Journal of Explosives Engineering* 14(5): 12–14.

Grant, G.R. & Dutton, A.J. 1983. Development of a fragmentation monitoring system for evaluating open stope blast performance at Mount Isa Mines. *Proc. 1st Int. Symposium on Rock Fragmentation by Blasting*, pp. 637–652.

Greenland, B.J. & Knowles, J.D. 1969. Rock breakage. *Mining Magazine* 120: 76–83.

Hagan, T.N. 1973. Good delay timing—prerequisite for effective bench blasts. *Proc. Australian Inst. Min. Met. No. 263*, pp. 47–54.

Hanspal, S., Scoble, M. & Lizotte, Y. 1995. Anatomy of a blast muck pile and its influence on loading machine performance. *Proc. 21st Conference on Explosives and Blasting Technique*, pp. 57–67.

Harries, G. 1987. The calculation of heave and muck pile profile. *Proc. 2nd Int. Symp. Rock Fragmentation by Blasting. Colorado*, pp. 248–256.

Hustrulid, W. 1999. Blasting principles for open pit mining. Vol. 1. Rotterdam: Balkema.

Kanchibotla, S. 2001. Optimum blasting? Is it minimum cost per broken rock or maximum value per broken rock. *Proc. Explo 2001, Hunter Valley, NSW, Australia.*

Katsabanis, P.D., Kelebek, S., Pelley, C. & Pollanen, M. 2003. Blasting effects on the grindability of rocks. *Proc. 29th Conf. on Explosives and Blasting Technique, Nashville, TN.*

La Rosa, D. 2001. The development of an information management system for the improvement of drilling and blasting in mining operations. *Proc. 29th International Symposium on Computer Applications in the Mineral Industries (APCOM), Beijing, China*, pp. 367–372.

Liu, Q. & Tran, H. 1996. Comparing systems—validation of Fragscan, Wipfrag and Split. *Proc. Measurement of Blast Fragmentation*. Rotterdam: Balkema, pp. 151–155.

MacKenzie, A.S. 1966. Cost of explosives—Do you evaluate it properly. *Mining Congress Journal*, 52(5): 32–41.

McGill, M. & Freadrich, J. 1994. The effect of fragmentation on loader productivity. *Proc. 5th State of the Art Seminar on Basting Technology*, pp. 713–724.

Nielsen, K. 1985. Sensitivity analysis for optimum open pit blasting. *Proc. 11th Conf. on Explosives and Blasting*, pp. 85–95.

Nielsen, K. 1987. Model studies of loading capacity as a function of fragmentation from blasting. *Proc. 3rd Mini Symposium on Explosives and Blasting Research*, pp. 71–80.

Ouchterlony, F., Nyberg, U. & Olsson, M. 2003. The energy balance of production blasts at Norkalk's Klinthagen quarry. *Proc. of the Second World Conference on Explosives and Blasting, Prague*, pp. 193–203.

Paley, N. & Kojovic, T. 2001. Adjusting blasting to increase SAG mill through put at the reddog mine. *Proc. 27th Annual conference on Explosives and Blasting Research, Orlando, FL.*

Sastry, V.R. & Chandar, K.R. 2004. Shocktube initiation for better fragmentation: A case study. *FRAGBLAST—Int. J. for Blasting and Fragmentation* 8(4): 207–220.

Simangunsong, G.M., Moser, P., Kramadibrata, S., Kubota, S., Shimada, H., Matsui, K. & Ogata, Y. 2003. Optimization of blasting fragmentation based on optimized comminution. *Proc. Annual Meeting, Japan Explosives Society, Tokio, May 15–16.*

Singh, S.P. & Narendrula, R. 2006. Factors affecting the productivity of loaders in surface mines. *International Journal of Mining, Reclamation and Environment* 20(1): 20–32.

Spathis. A.T. 1999. On the energy efficiency of blasting. Proc. 6th Int. Symp. on Rock Fragmentation by Blasting—FRAGBLAST-6, SAIMM, pp. 81–90.

Workman, L. & Eloranta, J. 2003. The effects of blasting on crushing and grinding efficiency and energy consumption. *Proc. 29th Conf. on Explosives and Blasting Technique, Nashville, TN*, 1, pp. 131–140.

Measurement and Analysis of Blast Fragmentation – Sanchidrián & Singh (Eds)
© 2013 Taylor & Francis Group, London, ISBN 978-0-415-62140-3

Return on experience from full-scale open pit blasting experiments

F. Delille
Expertise & Technical Services Department, AREVA MINES, Paris, France

D. Goetz & B. Tessier
Geosciences Department, MINES ParisTech, Fontainebleau, France

ABSTRACT: This paper gives practical feedback from an ambitious full-scale blasting program financed by AREVA MINES in 2010. The 32 experimental blasts consisted in single-hole, dual-hole and one-row (max. 5 holes) blasts in order to propose a fragmentation prediction method on a hole-by-hole basis. The horizontally-layered sandstone was blasted with ANFO in 6" diameter holes with bench heights between 10 and 17 m. Muckpile fragmentation measurements were set up to this aim at AREVA MINES's SOMAÏR open pit mine (Niger) according to standards; 2D truckload pictures were taken, sampled and processed with MINES ParisTech's FragScan software. Our paper emphasizes how site specific conditions and a highly productive environment (around 130 kt rock moved every day) restrain experimental possibilities, and practical recommendations to future experimenters are accordingly proposed.

1 INTRODUCTION

1.1 *Muckpile fragmentation: Implication on mining operations and measurement challenges*

Bench blasting is a widespread technique in open pit mining operations. Previous literature has provided many a proof that blast fragmentation has significant technical and economical impacts on all downstream production processes (e.g. crushing and grinding). Mine sites therefore strive for efficient prediction methods of their blasting results; finding accurate, efficient and cost-effective tools to measure and monitor blasting results is a continuous challenge for the blasting community.

Empirical formulae such as the Kuz-Ram model (see e.g. Cunningham 2005 for equations and earlier references) or the more recent KCO model (Ouchterlony 2005), as well as their bimodal spin-offs, are readily available to industrials for experimental data fitting. The accuracy of functions most used for rock fragmentation description has been compared (Sanchidrián et al. 2009).

The main challenge for full-scale on-site muckpile fragmentation measurements lies in sampling issues (Chavez et al. 1996). Dumper trucks must be cautiously monitored; but most importantly, a digital image of muck rock (e.g. truck load, or direct muckpile picture) is a surface sample of a 3D object—whether the image is 2D or 3D. This sampling bias, inherent to the measurement method, can never be fully solved in a mining operation framework. Therefore, digital image analysis techniques and particle delineation algorithms—though they form an essential part of the measurement process—are only a part of what makes acquired fragmentation data accurate and reliable in the full scale.

1.2 *Framework of herein discussed fragmentation measurements*

The fragmentation measurements were made in the frame of a large experimental program. Its aim was to develop and validate a fragmentation prediction method based on actual blast pattern parameters (*e.g.* effective bench and drillhole geometries) and made on a hole-by-hole basis.

32 tests sought to shed some light on blasting results by a comparative study of single-hole and multiple-hole blasting; fragmentation measurements were a central part in this field research, but other measurements such as 3D bench face profiles before and after blasting were also used in order to evaluate breakage results. A typical single-hole blasting result is shown farther in the paper (Fig. 3). Previous single-hole blasting studies in full scale had been published (Rustan & Nie 1987, Bilgin 1991, Bilgin et al. 1993), but to the purpose of our study, they lacked a comparison with multiple-hole blasting results. Main outcomes from the present research are to be published in a dedicated paper.

1.3 *The FragScan digital image analysis tool*

FragScan is a system designed to assess grain (or block) size distribution at different levels of fragmentation processes in the mining and quarrying industries. Other fragmentation measurement software has been developed in the past decade and is available to the mining industry, such as WipFrag, Split and PowerSieve. The four aforementioned systems' efficiencies have been compared in a blind study, on a fixed picture set (Latham et al. 2003).

The FragScan system operates from digital gray level photos; it is based on image analysis techniques allowing fast and automatic processing. This system was developed at the MINES ParisTech/ARMINES "Centre de Géosciences" in the early 1990s as part of projects involving the evaluation of blast design efficiency (Cheimanoff et al. 1993, Chavez Baroni 1996).

Besides well-identified errors related to digital extraction of size distribution of fragments visible on each image (Maerz & Zhou 1998)—unresolved finer fragments, overlapping fragments, over segmentation, image distortion ...—one fundamental problem remains the representativeness of the measure in relation with the studied object (Chavez et al. 1996).

Originally the system was designed to evaluate fragmentation of rock masses and the choice was to systematically grab pictures off the top of dumpers mucking rock fragments to the primary crusher (Schleifer & Tessier 1996). This procedure allowed on one hand to reduce the bias generated through sampling and on the other hand to minimize interruptions during production processes. Each image is not considered as individually representative of the dumper load but contributes to build an evaluation of overall muckpile block size distribution. Fragmentation monitoring was done by using CCTV cameras connected to a video card installed on a computer; running software allowed dumper regulation, image acquisition and image processing.

This method consisting in multiplying samples (Schleifer & Tessier 2002) has remained crucial in the implementation of the various versions of the system that emerged since its creation, whether for muckpile fragmentation applications or for online quality control of crushing and screening operations (raw material, riprap, aggregate ...). For the study achieved at SOMAIR, FragScan was adapted from the dumper version. The image acquisition system developed for SOMAÏR site with its specificities will be described in the next part.

2 EXPERIMENTAL CONDITIONS AND SETUP

2.1 *Mine and geology description*

AREVA MINES's open pit subsidiary SOMAÏR is located in Niger on the Western flanks of the Aïr mountains. SOMAÏR, at the time being, accounts for about 25% of AREVA's worldwide uranium production and has been operated since the early 1960's. Around 45 million tons of rock are blasted and mucked each year in the pits.

The mined geological layers were deposited in a fluviodeltaic environment. Overburden rock, in which blasting experiments were conducted, consists in a sub-horizontally alternated sequence of clayey sandstone layers, where clay-filled channels are locally present with a thickness varying between 10 and 50 cm. The rock mass is affected by no other systematic discontinuity set, though unsystematic discontinuities are present.

2.2 *Blasting experiments parameters*

As in the standard blasting practice on site in waste rock, hole diameter was 6" (152 mm). Blast-holes were loaded with bulk porous-prilled ANFO, which was mixed on site and loaded into holes by an explosives truck. Bottom hole priming was achieved by a 1 kg emulsion cartridge. Drill cuttings were used as stemming material; stemming height was 3.5 m. Electric detonators were used.

Holes were inclined by 15°. The aimed sub-drilling height was 1 m. Actual mean burden was the main varied parameter in single-hole tests (2.8–5.5 m). In dual-hole and one-row blasts, multiple burden/spacing configurations were tested. Bench heights varied between 10.1 and 17.1 m throughout the test program.

2.3 *Measurements other than fragmentation*

Actual bench face and breakage profiles after blasting were surveyed by stereophotogrammetry (BlastMetriX3D). Through this process, the volume torn off in blasts could be assessed and used for actual specific charge estimation.

Hole profiles were assessed by means of a Pulsar MK3 deviation probe. Hole collar positions were measured either in relative or in GPS coordinates with comparable accuracy. The combined use of bench face surveys, hole deviation and collar positions in a common coordinates system enabled actual burden value measurement for the holes.

Velocity of detonation (MREL DataTrapII recorder), explosives mass and density were also controlled in tests.

2.4 *Truckload pictures recording system*

A fixed recording installation where trucks would halt and be photographed was a too costly and cumbersome solution for the mine; as a consequence of site constraints which shall be detailed further in the next section, it was also impossible to take advantage of fixed installations such as the primary crusher unloading bay. Therefore, measurements had to be made without stopping or deviating trucks from their hauling path.

Dumper traffic was recorded on a digital video recorder connected to a CCTV camera which filmed the top of dumpers exiting the pit. The camera could be installed and removed quickly on a mast. Moving the mast itself to another measurement point on the other hand required truck-assisted handling. Images were selected from the movies according to literature suggestions (see e.g. Sanchidrián et al. 2006) and post-processed by FragScan on a blast basis to get the final granulometry. As an illustration, Figure 1 shows a truck passing underneath the camera.

Figure 1. View of the image recording system near the pit exit.

Figure 2. Example of 688×568 black & white truckload picture obtained with the system.

The video-recording system to which the camera was connected could record short sequences upon motion detection, which in our experience proved most convenient to reduce experimental data processing times. When recording images, a lossless compression recording format was chosen. Scaling of the 688×568 pixels pictures was done by indicating bucket length, which was measured prior to experiments (6500 mm). The fines cut-off size achieved by the system in its operating conditions was 40 mm. Figure 2 shows a typical picture obtained with the recording system.

3 RETURN ON EXPERIENCE

3.1 *Ore and waste blasting practices on site: Implications on fragmentation measurements*

There were a few site specific constraints at the SOMAÏR mine which imposed restrictions to blasting experiments possibilities, especially on the subject of fragmentation measurement.

The impact of sampling issues on fragmentation measurement accuracy is known to progressively decrease as the rock moves from the muckpile (e.g. muckpile photos) to the milling process (on-line fragmentation measurements in the primary crusher unloading bay for blasted rock, or on conveyor belts for crushed rock). Having several measurements along the hauling cycle (e.g. one measurement conducted on mining trucks, the other at the mill) is also preferable in order to increase confidence in measured data. Finally, when scalping is conducted before primary crushing, there are possibilities to measure the weight percentage of fragments smaller than the grizzly mesh size; this offers a calibration point to conduct fines corrections on the raw particle size distributions obtained by digital image analysis. For all these reasons, it would be generally preferable to conduct blasting experiments on ore blasts rather than on waste blasts, where rock is directly hauled to the waste dump.

Now, in the SOMAÏR case, the radioactive nature of uranium ore enables selective mining. Grade control is conducted for all ore blasts, with the help of radioactivity probes, and the ore blasting practice is designed to achieve maximum mining accuracy: reduced height benches are blasted (6 m), small diameter holes (89 mm) are used in order to increase grade control resolution, and heave blasting is used.

Bench faces in ore are therefore always blocked by the previous blast's muckpile, making it impossible to measure actual bench face profiles before blasting—and as a consequence to establish a pertinent value for blasthole burden with classical techniques.

Furthermore, SOMAÏR conducts blending for the ore mix sent to the mill: several temporary ore stockpiles are made, with a classification according to ore grade and clay contents. Ore trucks do not directly send their load to the crusher but dump into these stockpiles. This breaks the hauling cycle and makes it impossible to link a picture taken at the primary crusher bay to a given blast in the pit.

For these two main reasons, fragmentation measurements in our experiments were conducted in waste blasts. In these conditions, any method to assess an actual weight percentage of fragments below the fines cut-off size of the digital image sampling device would have been prohibitive in terms of costs.

The few site specific constraints we have evoked here, though they may not seem relevant from a preliminary perspective, impose tremendous restrictions on blast fragmentation measurement possibilities.

3.2 *Sharing means with a production site*

SOMAÏR mine very generously allocated all necessary staff and machines while we tried to keep their use in experiments to a minimum. Mine management and crews put up a huge effort to support the experimental program.

Despite all efforts, however, there was a significant gap between planned and actual experiment durations. Due to production constraints and/or mechanical problems, specific machines (e.g. drills, wheel loaders and trucks, diesel power generator) or mine teams (e.g. surveyors for GPS measurements, geologists for qualitative bench face analysis, shotfirer crews) often happen to be busy or unavailable for the experimental task required.

The multiplicity of resources involved increases this effect which must not be overseen in experimental program timing and planning. Our experience on this test program dictates that in similar site conditions, increasing theoretical experimental durations by 50 to 100% will give a reasonable estimate of actual times needed.

In relation to this aspect, we also suggest that a minimum of two persons working full time to conduct and monitor blasting experiments should be considered: this enables for one experimenter to permanently be on the field, conducting or monitoring data acquisition, while the other may address logistical issues and analyze data at the mine offices—in our case, distances and speed limits were such that half an hour was needed to only travel back and forth from the pit to the offices.

3.3 *Constraints imposed by mine planning method*

Many a paper has emphasized that the greatest difficulty in full scale conditions is to achieve sufficient repeatability in experiments, *i.e.* to keep any parameter which is not studied in the test program as constant as possible. This can never be achieved to a full extent, be it only because of geological variations from one spot to another.

The best practical solution to achieve maximal repeatability would be to conduct all experiments in a fixed perimeter.

SOMAÏR mine planning is based on simultaneously progressing horizontal benches; this planning method offers increased flexibility, but pit evolution may therefore easily change in the medium term—in contrast to the pushback planning method. Changes in short and medium term pit planning constraints made it impossible to keep an assigned experimental perimeter since the experiments planning spread over several months.

All experiments however remained at the same bench level in order to conduct them in a fixed geological layer.

3.4 *Minimum spacing between blasts*

In our experiments, keeping test bench widths as small as possible and making as many experiments in a test bench as possible was beneficial for various reasons. The most important one is that a full blasting experiment required around a full week from its beginning to its end; this was impractical for the mine, because a full area of the pit was mobilized during that time. It was therefore intended, in a test bench, to keep the spacing between adjacent experiments as small as possible.

In the first test, spacing between experiments was insufficient because breakage span was underestimated; a safety distance of 6 burdens between adjacent experiments was applied afterwards with satisfactory results (see Fig. 3).

3.5 *Issues related to trucks sampling*

From the experience gained in using the truckload images sampling device at SOMAÏR, two interesting points may be noted.

Figure 3. Single hole tests R3T3 and R3T4. Mean bench height 16.5 m. Actual, mean hole burdens 4.6 and 3.6 m respectively.

Since the Less Fines Project, the use of mine fleet management systems has been spreading in the mining industry. However, not all mine sites are equipped with systems complex enough to fit the experimenters' needs. In our case, the SOMAÏR mine was equipped with a "passive" MFMS, which did not allow for the reactive control and redirection of dumper trucks which is essential in correct sampling. In site conditions where a reactive MFMS is available, light adaptations to the system would provide significant help to the experimenter.

At SOMAÏR, we used flags which allowed quick attachment and removal, and easy identification of trucks on pictures; be that as it may, the mine site size required a constant vigilance for many reasons.

A pit often has more than one access slope (in our case, for most blasting experiments, two were simultaneously active); truck drivers might therefore take the wrong one and unexpectedly circumvent the fragmentation measurement installation. A mining supervisor may also be in urgent need of an additional truck to work on another muckpile and request some of the experimental trucks to come over. If the drivers still take same access slope and the experimenter is unaware of this change, truckload pictures will be sampled which do not come from the experimental blast. There is also the reverse possibility that a mining supervisor will send over new trucks which are unflagged, resulting in truckload loss in the sampling process.

All these situations were observed at least once but could be quickly corrected thanks to uninterrupted presence and monitoring during mucking operations, which is a very time-costly process.

The second observation is related to motion detection recording. Video recording was triggered as soon as motion was detected on camera, and the system kept on recording until movement stopped again. When downloading the videos, the system showed the first picture of the sequence, which enabled to quickly identify videos triggered by an experimental truck passing.

In a limited number of cases however, several truckloads including an experimental truck gathered into one same recording, without the triggering vehicle being an experimental truck. Corresponding pictures were thought to be lost at first. As soon as the issue was well known, it was possible to retrieve most of the previously "lost" pictures by back-analysis of older raw data on the recorder.

3.6 *Powering the image acquisition system*

The pit access slope (hence the image recording system) was several hundreds of meters away from the 330 V mine electricity network. Powering the system through said network would have required deploying a specific cable as well as earth works to bury all parts where the cable would intersect roads, then a transformer to provide a stable 220 V output.

Instead, a diesel power group readily equipped with 220 V output was used, which served during the night for local backup lighting in the pit. A solar power solution had also been considered but was too costly for the sole experiments' purpose.

The use of a diesel power generator had some drawbacks: first, one is dependent on mine operations to start mucking (teams towing power groups may have to work on other priorities and delay bringing it, or the refuelling team may also be unavailable); second and most importantly, there is no noticing if there is an unexpected power loss.

The case occurred in an experiment and caused the loss of three out of four muckpile fragmentation curves because of insufficient sampling (the power loss occurred early and less than 20% of the truckload pictures were actually taken). In all subsequent tests, regular checks to the power group were imposed to circumvent this risk.

3.7 *Influence of fines corrections on fragment size distribution results*

Since a reliable measurement of fines mass percentages in muckpiles could not be made, we resorted in our experimental data analysis to using the raw fragmentation curves obtained with the system. Such fragmentation results can only be considered as a local indicator to compare experimental results on a same site.

Be that as it may, to evaluate the impact of fines calibration on fragmentation indicators such as X_{50} and n obtained in the study, we conducted a simplified fines correction of our experimental curves. The "optical" fines percentage—which is the ratio of visible fine patches to total truckload pixels on a given image—was used as an input.

The correction was based on following simplifications: a) surface measurements on truckloads reflect the fragment size properties of material hidden underneath; b) height of the mucked loads is homogeneous in the dumper body and is 1 m; c) fragment shapes are spherical; d) all fragments in a [X1 – X2] sieve have an equivalent diameter of (X1 + X2)/2.

After correction, Rosin-Rammler uniformity indexes of curves dropped between 0.50 and 0.75 (against 1.20 to 1.80 in the uncorrected case). Mean fragment sizes dropped between 35–470 mm (against 510 to 1680 mm in the uncorrected case). Figure 4 shows the example of mean fragment sizes in the corrected *vs.* uncorrected case.

As a consequence, q-exponents and prefactors in Kuznetsov-like empirical laws between mean

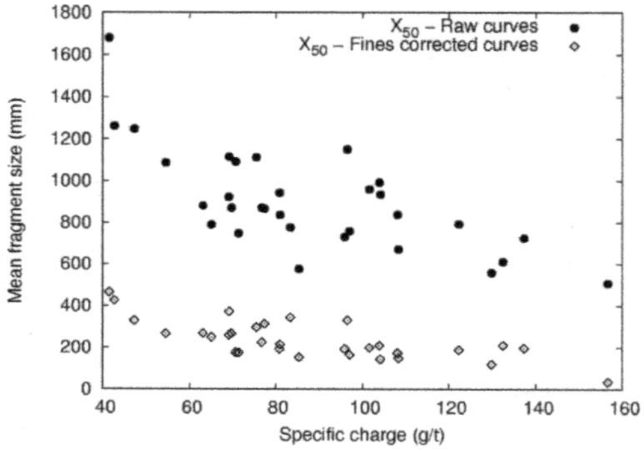

Figure 4. Mean fragment sizes obtained in blasting experiments as a function of specific charge—uncorrected and fines-corrected.

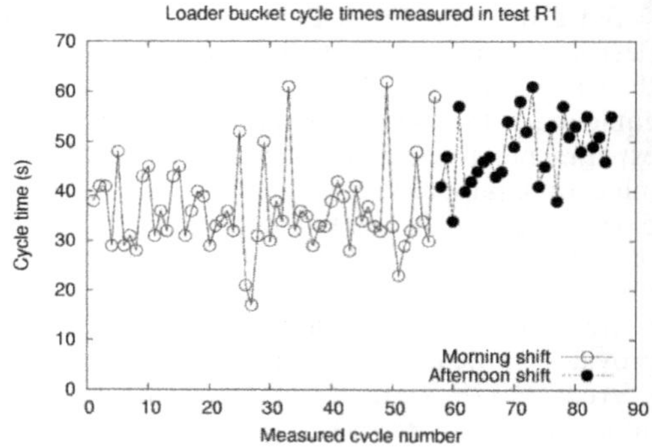

Figure 5. Recorded wheel loader bucket cycle times during the first test in the experimental program (R1). Colour change in plotted dots indicates shift change.

fragment size and specific charge are also significantly impacted. In the example of one-row blasts, comprising four or five holes with 25 ms delays between each, the q-exponent in the best-fit $X_{50}(q)$ power law increased from 0.76 (uncorrected) to 1.73 (corrected).

3.8 *Repeatability issues in productivity measurements*

One of the initial plans in the test program was to monitor wheel loader buckets loading cycle times and to seek a link between said cycle time and muck-pile fragmentation (e.g. mean fragment size X_{50}).

The influence of driver performance on cycle times has already been pointed out several times in past literature; one of the observations made in our tests confirms this idea and suggests that human effects on loading equipment performance probably outscore fragmentation-linked effects.

Figure 5 shows a chronological plot of measured cycle times in seconds. The sudden leap in cycle times is simultaneous with the afternoon shift change. Now, repeatability may be increased by selecting a driver and using the same loader every time; however, this becomes very difficult to achieve on a large scale mine site where teams rotate and where many loaders are used.

3.9 *Getting insightful information out of quick measurements*

Among other fragmentation results in our experimental study, there is a good correlation between X_{max} and raw $X_{50,}$ both in single-hole and delayed, multiple-hole blasting conditions; furthermore, the dependence of X_{max} on X_{50} is significantly different in both cases. These parameters are not correlated in the simultaneously initiated dual-hole case though.

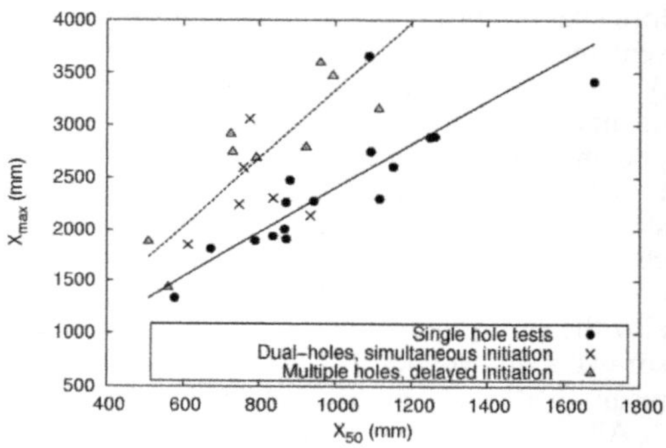

Figure 6. Comparison between X_{max} and X_{50} for different test types.

Regressions (1) and (2) for the single-hole and delayed multiple-hole cases respectively are provided below. Corresponding data points and correlations are plotted on Figure 6.

$$X_{max}^{SH} = 5.59 \cdot X_{50}{}^{0.88} (mm);\ R^2 = 0.78 \qquad (1)$$

$$X_{max}^{\Delta T} = 4.05 \cdot X_{50}{}^{0.97} (mm);\ R^2 = 0.74 \qquad (2)$$

In other words, in our experimental conditions, X_{50} and X_{max} measures are equivalent to a certain extent. Let it be noted that this result is likely to be site specific: raw X_{50} values have been taken into account, because as previously explained no fines correction of the measured fragmentation curves was possible.

From an operational point of view on site, this result is far from negligible. Blast fragmentation in the coarse domain would be monitored to an industrially acceptable approximation by only observing blocks passing on trucks and discarding the cumbersome image processing phase required in thorough digital fragmentation analysis.

If reliable automatic detection algorithms can be used to distinguish blocks from fines patches (in order to discard the latter), a statistically meaningful online evaluation of fragmentation intensity in blasts will probably be obtained. As of today, such a method could not be envisaged at SOMAÏR though, because a full computer assisted mine fleet management system structure is a prerequisite.

The suggestion of monitoring block sizes "on the flow" does not suppress the need for careful truck sampling—to make sure a recording can reliably be traced back to the right muckpile, hence to the right blasting parameters.

It would be interesting to investigate whether similar results can be witnessed on other sites; previous literature already suggested that, apart from other influential factors such as rock mass jointing, X_{50} and X_{max} might in general follow resembling laws (Ouchterlony 2010).

4 CONCLUSION

Most points emphasized in this paper join into the conclusion that the size of large mine sites, and productivity stresses usually associated to them, pose practical difficulties which significantly increase the effort needed to conduct a repeatable blasting experiments program. In many ways, experimenting in medium size quarries is preferable. Remarks made are applicable to most other extraction cycle processes both open pit and underground, because the main issue is linked to time and physical means required to execute experimental tasks. An experiment may also stand on the critical path to next production targets. This is particularly the case in open pit, where freezing progression on the upper bench creates a constraint for progression of benches underneath.

Generally speaking, the issue of whether the blasted rock will be milled or not does not apply in a quarry where a great majority of the blasted material is sent to the crusher. In this context, more fragmentation measurement solutions are available to the experimenter.

ACKNOWLEDGEMENTS

Authors wish to thank AREVA MINES for initiating this research work and providing full financial support for its conduction; they warmly thank the SOMAÏR mine site for accepting to host the experimental program and provide all necessary mine equipment and staff for its realization, as well as for never withdrawing its support even when especially difficult production issues were to be faced.

Mine engineer Moussa Ousmane, who was put in charge of conducting almost half of the blasting experiments and consistently helped in leading all others, is especially thanked; we feel indebted to his continued efforts. All other SOMAÏR mine staff who have collaborated to the field work are thanked for their implication.

REFERENCES

Bilgin, H.A. 1991. Single hole test blasting at an open pit mine in full scale: a case study. *Int. Journal of Surface Mining and Reclamation* 5: 191–194.

Bilgin, H.A. Pasamehmetoglu, A.G. & Ozkahraman, H.T. 1993. Optimum burden determination and fragmentation evaluation by full scale slab blasting. *Proc. 4th Int. Symposium on Rock Fragmentation by Blasting*, pp. 337–344.

Chavez Baroni, R.N. 1996. Mise au point d'outils pour le contrôle du tir à l'explosif sous contraintes de production. Ph.D. thesis, Paris School of Mines.

Chavez, R., Cheimanoff, N. & Schleifer, J. 1996. Sampling problems during grain size distribution measurements. *Proc. 5th Int. Symposium on Rock Fragmentation by Blasting*, pp. 245–252.

Cheimanoff, N., Chavez, R. & Schleifer, J. 1993. FRAGSCAN: A scanning tool for fragmentation after blasting. In H.P. Rossmanith (ed.), *Rock Fragmentation by Blasting: Proceedings of the Fourth International Symposium on Rock Fragmentation by Blasting, FRAGBLAST-4*, pp. 325–329.

Cunningham, C.V.B. 2005. The Kuz-Ram fragmentation model—20 years on. *Proc. 3rd EFEE Conference*, pp. 201–210.

Latham, J-P., Kemeny, J., Maerz, N., Noy, M., Schleifer, J., & Tose, S. 2003. A blind comparison between results of four image analysis systems using a photo-library of piles of sieved fragment. *Fragblast—International Journal for Blasting and Fragmentation* 7(2): 105–132.

Maerz, N.H. & Zhou, W. 1998. Optical digital fragmentation measuring systems—inherent sources of error. *Fragblast—International Journal for Blasting and Fragmentation* 2(4): 415–431.

Ouchterlony, F. 2005. What does the fragment size distribution of blasted rock look like? *Proc. 3rd EFEE Conference*, pp. 189–199.

Ouchterlony, F. 2010. A common form for fragment size distributions from blasting and a derivation of a generalized Kuznetsov's x_{50}—equation. *Proc. 9th Int. Symposium on Rock Fragmentation by Blasting*, pp. 199–208.

Rustan, A. & Nie, S.L. 1987. New method to test the rock breaking properties of explosives in full-scale. *Proc. 2nd Int. Symposium on Rock Fragmentation by Blasting*, pp. 36–47.

Sanchidrián, J.A., Segarra, P. & López, L.M. 2006. A practical procedure for the measurement of fragmentation by blasting by digital image analysis. *Rock Mechanics and Rock Engineering* 34(9): 359–382.

Sanchidrián, J.A., Segarra, P., López, L.M., Ouchterlony, F. & Moser, P. 2009. Evaluation of some distribution functions for describing rock fragmentation data. *Proc. 9th Int. Symposium on Rock Fragmentation by Blasting,* pp. 239–248.

Schleifer, J. & Tessier, B. 1996. FRAGSCAN: A tool to measure fragmentation of blasted rock. In J.A. Franklin & T. Katsabanis (eds.), *Measurement of Blast Fragmentation: Proceedings of the FRAGBLAST 5 Workshop on Measurement of Blast Fragmentation*, pp. 73–78.

Schleifer, J. & Tessier, B. 2002. Fragmentation Assessment using the FragScan System: Quality of a Blast. *Fragblast: International Journal for Blasting and Fragmentation* 6(3–4): 321–331.

Measurement and Analysis of Blast Fragmentation – Sanchidrián & Singh (Eds)
© 2013 Taylor & Francis Group, London, ISBN 978-0-415-62140-3

Fragmentation risk assessment in open-pit blasting using interaction matrix-vulnerability index method

F. Faramarzi, H. Mansouri & M.A. Ebrahimi Farsangi
Mining Engineering Department, Shahid Bahonar University of Kerman, Kerman, Iran

M. Monjezi
Faculty of Engineering, Tarbiat Modares University, Tehran, Iran

ABSTRACT: The purpose of blasting in open-pit mines is achieving a desired fragment size distribution, considering safety and environmental aspects. Regarding economic aspects, fragmentation plays a critical role in the total costs of mining operation. Poor fragmentation, oversize and undersize, means economic loss and as a consequence, putting the whole mining operation at risk. Risk assessment prior to blasting operation can be a helpful tool for total cost reduction. This paper presents a new model to analyze the risk due to poor fragmentation in rock blasting, based on the concept of Rock Engineering Systems (RES). The principles of RES were used to define an index, the so called vulnerability index (VI), expressing the risk condition due to poor fragmentation. In defining the new model, two main steps must be taken into account. The first step is to identify the parameters that are responsible for the occurrence of risk in the case of fragmentation (poor fragmentation), analyze their behavior and evaluate the significance (weight) that each one has in the overall risk conditions (poor fragmentation). In this step, the RES principles can be used to assess the weighting of the parameters involved. In the second step, the vulnerability index can be determined, which shows the level of risk expected for each blast. The data for 30 blasts, carried out in Sungun copper mine, western Iran, were used to validate the model. The results obtained indicating the capability of the model proposed.

1 INTRODUCTION

Blast design combines the aspects of safety, cost efficiency and the control of environmental effects. Being aware of the forthcoming hazards before starting any project and estimating risk levels can prevent damages and undesirable effects, excessive financial costs, and also operation stoppage. Rock fragmentation has been the concern of many research works (Mackenzie 1967, Latham et al., 1999, Sanchidrian et al., 2007) because it is considered as the most important aspect of production blasting, since it affects on the costs of drilling and blasting and the efficiency of all the subsystems such as loading, hauling and crushing in mining operations (Mackenzie 1967).

The parameters affecting on the rock fragment size and the corresponding risk level can be categorized in two groups: the first group is controllable parameters; such as blasting design parameters and also explosive related parameters; and the second one are uncontrollable parameters, which contains physical and geomechanical properties of intact rock and also rock mass (Singh & Sastry 1986, Gosh et al., 1990, Kulatilake et al., 2010).

Poor fragmentation, oversize and undersize, means economic loss and as a consequence, putting the whole mining operation at risk. Risk assessment prior to blasting operation can be a helpful tool for total cost reduction. In this paper, a new model, considering all pertinent parameters is proposed to predict the risk due to poor fragmentation in rock blasting, using the concept of Rock Engineering Systems (RES) with interaction matrix. The principles of RES were used to define an index, the so called vulnerability index (VI), describing the risk condition due to poor fragmentation.

The RES approach has been applied to a number of rock engineering fields, for examples, evaluation of stability of underground excavations (Lu & Hudson 1993), hazard and risk assessment of rockfall (Cancelli & Crosta 1993), rock mass characterization for indicating natural slope instability (Mazzoccola & Hudson 1996), development of an assessment system for blastability of rock masses (Latham & Lu 1999), assessing geotechnical hazards for TBM tunnelling (Benardos & Kaliampakos 2004) and quantitative hazard assessment for tunnel collapses (Shin et al., 2009).

To validate the performance of the model proposed, it was applied to Sungun copper mine, Iran.

2 DEVELOPED METHOD FOR RISK ASSESSMENT OF FRAGMENTATION

2.1 *Rock engineering systems*

The concept of RES, introduced by Hudson (1992), is a method of structuring all the ways in which rock mechanics parameters and variables can affect one another-the rock mechanics interactions. The key element in the RES is interaction matrix. The interaction matrix is both the basic analytical and a presentational technique for characterizing the important parameters and the interaction mechanisms in a rock engineering system. The generation of the interaction matrix can help in evaluating the weighting of the parameters within the rock mass system as a whole.

In the interaction matrix, the principal parameters affecting the system (for example, a fragmentation system) are located along the leading diagonal of the matrix and the effects of each individual parameter on any other parameter (interactions) are placed on the off-diagonal cells. The assigning values to off-diagonal cells are called coding the matrix. A problem having only two parameters is the simplest of interaction matrix, as revealed in Figure 1a. Also, a general view of the coding of interaction matrix is shown in Figure 1b (After Hudson 1992). The row passing through P_i represents the influence of P_i on all the other parameters in the system, while the column through P_i shows the effects of other parameters, or the remaining of the system, on the P_i. In principal, there is no limit to the number of parameters that may be included in an interaction matrix.

Different procedures such as, the 0–1 binary, expert semi-quantitative (ESQ) (Hudson 1992) and the continuous quantitative coding (CQC) (Lu & Latham 1994) were proposed for numerically coding the interaction matrix. Among the coding procedures, the ESQ coding is the most common used. According to this coding technique, the interaction intensity is denoted by the values from 0 (no interaction) to 4 (critical interaction) as shown in Table 1.

In the interaction matrix, the sum of a row is called the "cause" value $\left(E_{Pj} = \sum_{i=1}^{n} I_{ij}\right)$ and the sum of a column is the "effect" value $\left(C_{Pi} = \sum_{j=1}^{n} I_{ij}\right)$ denoted as coordinates (C, E) for a particular parameter. The coordinate values for each parameter can be plotted in cause and effect space, forming the so-called C – E plot.

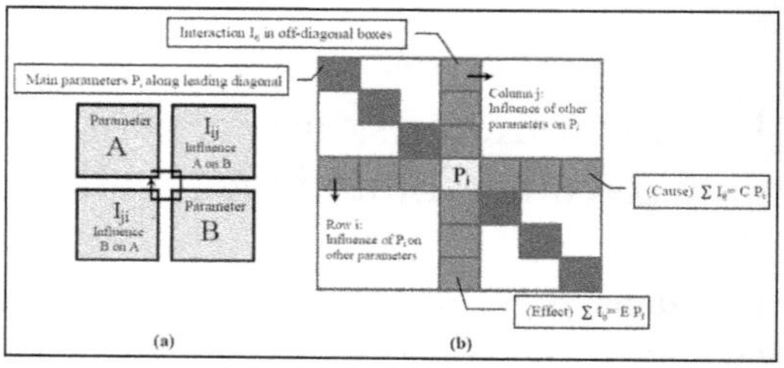

Figure 1. Interaction matrix in RES, a: two parameters interaction matrix; b: a general view of the coding of interaction matrix (After Hudson 1992).

Table 1. ESQ interaction matrix coding (Hudson 1992).

Coding	Description
0	No interaction
1	Weak interaction
2	Medium interaction
3	Strong interaction
4	Critical interaction

The interactive intensity value of each parameter is denoted as the sum of the C and E values (C + E) and it can be used as an indicator of parameter's significance in the system. The percentage value of (C + E) can be used as the parameter's weighting factor (a_i) as shown in Equation 1.

$$a_i = \frac{(C_i + E_i)}{\left(\sum_i C_i + \sum_i E_i\right)} \times 100 \qquad (1)$$

2.2 *Vulnerability index*

The principles of RES were used in the vulnerability index (VI) methodology concept, first introduced by Benardos et al., (2004) to identify the vulnerable areas that may pose threat to the Tunnel Boring Machine (TBM) tunneling operation. In this research, a similar methodology, inspired by the work carried out by Benardos et al., (2004) was adopted to define a model to predict risk level associated with the muck pile fragment size, considering poor fragmentation as risk encountered during blasting operation.

In assessing the risk of fragmentation, two main steps must be taken into account. The first step is to identify the parameters that are responsible for the occurrence of risk in case of fragmentation (poor fragmentation), analyze their behavior and evaluate the significance (weight) that each one has in the overall risk conditions (poor fragmentation). In this step, the RES principles can be used to assess the weighting of the parameters involved (within the rock mass system as a whole).

In the second step, the vulnerability index can be determined, using Equation 2 (Benardos et al., 2004):

$$VI = 100 - \sum_{i=1} a_i \frac{Q_i}{Q_{max}} \quad (2)$$

where; a_i is weighting of the ith parameter, Q_i is value (rating) of the ith parameter and Q_{max} is maximum value assigned for ith parameter (normalization factor).

Based upon the vulnerability index estimated (Equation 2) classification of the vulnerability index, which is divided into 3 main categories with different severity of the normalized scale of 0 to 100, (Table 2) (Benardos et al., 2004), the level of risk corresponding to fragmentation of a blast can be identified. In category I, small-scale problems are expected, that cannot significantly affect the results of fragmentation of the blast. In category II, the problematic occurrence of poor fragmentation might encountered, which must be taken into account. In category III, certain individual regions with poor fragmentation, which might cause several difficulties during the loading, hauling and crushing must be considered.

2.3 *Parameters influencing on the muck pile fragment size*

In reviewing the literatures published, addressing the parameters affecting on the muck pile fragment size (Langefors & Kihlstrom 1978, Jimeno et al., 1995, Bhandari 1997, Hustrulid 1999), it is clear that many parameters can influence on the muck pile fragment size and the corresponding risk level. However, the 16 parameters given in Table 3 represents the most important ones and also the most easily obtainable without having to resort to particular site investigation techniques. Blastablility Index (BI) in the Table 3 represents the effects of rock structure on fragmentation.

2.4 *Interaction matrix and rating of parameters*

The 16 principal parameters affecting on the muck pile fragment size are located along the leading diagonal of the matrix and the effects of each individual parameter on any other parameter (interactions) are placed on the off-diagonal cells. The assigning values to off-diagonal cells, coding the matrix, were carried out, using the ESQ coding method as proposed by Hudson (1992). Based upon the views of 3 experts, the interaction matrix for the parameters affecting on the muck pile fragment size is established as presented in Table 4. Table 5 gives cause (C), effect (E), interactive intensity (C + E), dominance (C − E) and weight of each parameter (a_i) for fragmentation purposes. As it can be seen in Table 5, burden, BI, H/B ratio and powder factor have the highest weights of 11.7%, 9.4%, 8.3% and 8% respectively, which in comparison with other parameters highly control the whole fragmentation system.

E–C histogram and C + E for each parameter are illustrated in Figures 2 and 3 respectively. The points below the C = E line are called dominant and the points above the C = E line are called subordinate.

Table 2. Classification of the vulnerability index (Modified after Benardos et al., 2004).

Risk description	Low-medium	Medium-high	High-very high
Category VI	I 0–33	II 33–66	III 66–100

Table 3. Effective parameters on the muck pile fragment size.

Effective parameters			
P_1	Burden (B)	P_9	Blasthole inclination
P_2	Maximum instantaneous charge	P_{10}	Blasthole deviation
P_3	Powder factor	P_{11}	Hole diameter (D)
P_4	S/B ratio (S: spacing)	P_{12}	J/B ratio (J: subdrilling)
P_5	S_T/B ratio (S_T: stemming)	P_{13}	Blast hole pattern
P_6	Stiffness ratio (H/B) (H: height of bench	P_{14}	Initiation sequence
P_7	Number of rows	P_{15}	Blastability Index (BI)
P_8	Time delay	P_{16}	B/D ratio

Table 4. The interaction matrix for the parameters affecting on the muck pile fragment size.

P_1	2	2	2	2	2	2	3	2	0	0	2	0	0	0	0
0	P_2	0	0	0	0	0	1	0	0	0	0	0	0	0	0
2	0	P_3	2	0	0	0	0	0	0	0	0	0	0	0	2
0	2	2	P_4	0	0	0	0	0	0	0	0	0	0	0	0
0	2	1	0	P_5	0	0	0	0	0	0	0	0	0	0	0
2	2	2	3	2	P_6	1	1	1	2	2	2	0	0	0	1
1	1	1	0	1	0	P_7	3	0	0	0	0	0	0	0	0
0	1	0	0	0	0	1	P_8	0	0	0	0	0	0	0	0
2	2	2	0	1	0	0	1	P_9	2	0	2	0	0	0	0
2	1	1	2	1	0	0	0	0	P_{10}	0	1	0	0	0	1
3	3	2	1	1	2	1	2	0	3	P_{11}	0	0	1	0	2
0	1	1	0	0	0	0	0	0	0	0	P_{12}	0	0	0	0
2	0	2	2	0	0	1	0	0	0	0	0	P_{13}	2	0	1
2	2	0	1	0	0	2	2	0	0	0	0	0	P_{14}	0	1
3	2	4	2	1	2	2	3	2	3	1	2	2	1	P_{15}	3
3	1	2	0	0	2	1	3	0	0	0	1	1	1	0	P_{16}

Table 5. Weighting of the principal parameters in rock fragmentation.

Parameter	C	E	C + E	C − E	a_i (%)
P_1	19	22	41	−3	11.7
P_2	1	22	23	−21	6.6
P_3	6	22	28	−16	8.0
P_4	4	15	19	−11	5.4
P_5	3	9	12	−6	3.4
P_6	21	8	29	13	8.3
P_7	7	11	18	−4	5.1
P_8	2	19	21	−17	6.0
P_9	12	5	17	7	4.9
P_{10}	9	10	19	−1	5.4
P_{11}	21	3	24	18	6.9
P_{12}	2	10	12	−8	3.4
P_{13}	10	3	13	7	3.7
P_{14}	10	5	15	5	4.3
P_{15}	33	0	33	33	9.4
P_{16}	15	11	26	4	7.4
Sum	175	175	350	0	100

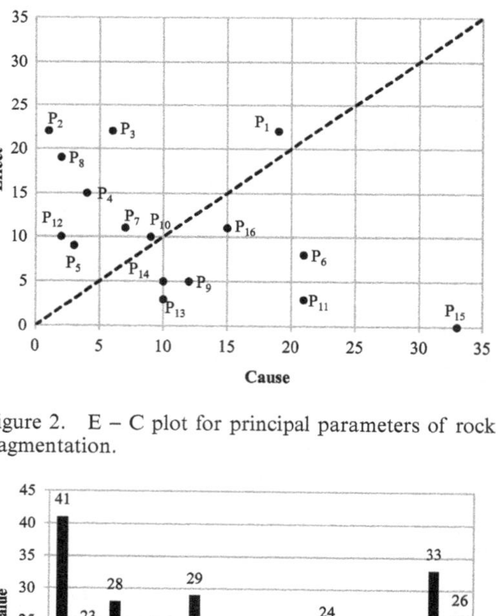

Figure 2. E − C plot for principal parameters of rock fragmentation.

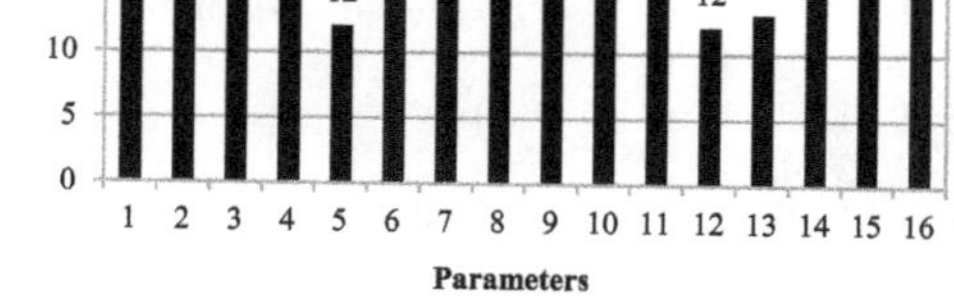

Figure 3. The C + E values for principal parameters of rock fragmentation.

2.5 *Rating of parameters*

The rating of the parameter's values was carried out based upon their effect on the muck pile fragment size. Totally 6 classes of rating, from 0 to 5 were considered, where 0 denotes the worst case (most unfavorable condition as poor fragmentation with high risk) and 5 the best (most favorable condition as good fragmentation with low risk). In the case of rock fragmentation, the rating of each parameter is presented in Table 6. The ranges of parameters in Table 6 were proposed based on the empirical results, practical limits and the experiences of different researchers.

3 SITE DESCRIPTION

3.1 *Location and geological properties*

Sungun, an open-pit copper mine with a mineable reserve of 410 Mt, is planned to produce 7 Mt ore for the initial 7 years with the intention to expand capacity up to 14 Mt ore with an average grade of 0.6 copper; is located 100 km north east of Tabriz city, Iran (Figure 4).

The Sungun intrusive complex hosting the Sungun porphyry copper stock intruded along the Sungun anticline into cretaceous limestone, marls and shales. The main lithological units exposed in the Sungun pit are Sungun porphyry, Dykes, Pyroclastics, Trachybasalt and Skarn. Also, UCS differs from 55 to 82 MPa in different lithological units of mine. The initial design of the final pit indicated a maximum slope height of 765 m.

A general view of the Sungun copper mine is shown in Figure 5. Also, the range of measured blasting parameters for 30 production blasts in this mine is given in Table 7.

4 RISK ANALYSIS OF FRAGMENTATION BASED ON THE METHOD PROPOSED

The data related to 30 production blasts, carried out at Sungun copper mine, were applied to determine the associated VI for each blast, using Equation 2. Table 8 presents the measured X_{80} which was determined by image analysis method and their associated VI for samples of 30 production blasts. Variations in the VI for the 30 blasts are shown in Figure 6. As it can be seen, VI varies from 35 to 58, showing that the level of risk is in the second category (Medium-High).

Also, there is a good agreement between VI and X_{80} for each blast, as shown in Figure 7. For instance, blast No. 7 has the minimum risk of 38 (out of 100) and the associated X_{80} is 10 cm. Also,

Table 6. Proposed ranges for the parameters effective in the fragmentation and associated risk level.

No.	Parameters		Values/description and ratings					
1	Burden (m)	value	<3		3–5	5–7	7–9	>9
		rating	4		3	2	1	0
2	Maximum Instantaneous charge (kg)	value	<500	500–1000	1000–2000	2000–3000	3000–4500	>4500
		rating	5	4	3	2	1	0
3	J/B ratio	value		<0.1	0.1–0.3	0.3–0.5		>0.5
		rating		0	1	3		2
4	S/B ratio	value	<1		1–2	2–3	3–4	>4
		rating	0		3	2	1	0
5	S_T/B ratio	value	<0.7		0.7–0.9	0.9–1.2	1.2–1.4	>1.4
		rating	0		2	4	3	1
6	Number of rows	value	<3		3–5	5–6	6–7	>7
		rating	4		3	2	1	0
7	Stiffness ratio (H/B)	value	<1		1–2	2–3	3–4	>4
		rating	0		1	2	3	4
8	Blast hole pattern	description	Staggered		square	Rectangular		Single row
		rating	3		2	1		0
9	Initiation sequence*	description	V_{34}		V_{64}	V_{34}	V_{45}	Inter row
		rating	4		3	2	1	0
10	Hole deviation (°)	value	0		0–5	5–10	10–15	>15
		rating	4		3	2	1	0
11	Blastability Index (BI)	value	0–20		21–40	41–60	61–80	81–100
		rating	4		3	2	1	0
12	Hole inclination (°)	value	90		90–80	80–70	70–65	65
		rating	0		1	2	3	2
13	Powder factor (g/ton)	value	<125	125–150	150–175	175–210	210–300	>300
		rating	0	1	2	3	4	4
14	Time delay (ms/m)	value	<2	2–5	5–7	7–10	10–20	>20
		rating	0	1	3	4	2	1
15	Hole diameter (mm)	value	<100	100–150	150–200	200–250	250–300	>300
		rating	4	3	2	1	0	0
16	B/D ratio	value			<20	20–40		>40
		rating			2	1		0

* V_α denotes the angle between crest and shot row, inter row means that the blast initiates row by row.

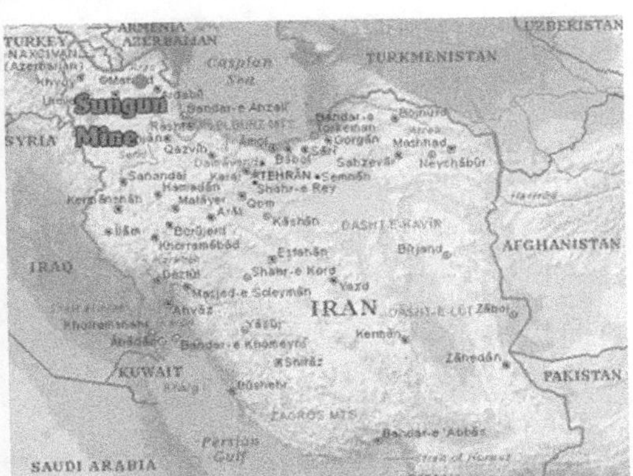

Figure 4. Location of Sungun copper mine.

Figure 5. A general view of the Sungun copper mine.

Table 7. Range of the measured blasting parameters in Sungun copper mine.

Parameter	Range
Burden (m)	1–4.8
Maximum instantaneous charge (kg)	81–3245
Powder factor (g/ton)	72–1010
S/B ratio	1–6.7
S_T/B ratio	0.33–3.44
Stiffness ratio	1.1–10
Number of rows	1–5
Time delay (ms)	20–200
Hole inclination (degree)	90
Hole diameter (mm)	125,150,162.5, 250
Hole deviation (degree)	10–15
B/D ratio	5–38
J/B ratio	0–0.4
Blast hole pattern	Different
Initiation sequence	Inter row
Blastability Index	33.5–63.5
Rock density (g/cm³)	2.3–2.5

Table 8. Samples of measured fragment size and their associated VI.

Blast no.	Blast location	X_{80} cm	VI	Date (YY/MM/DD)
1	2200-M	65	46	2010/12/13
3	1975-N	60	41	2010/12/13
5	1937.5-N	43	41	2010/12/14
7	2212.5-M	48	37	2010/12/19
9	2012.5-N	10	35	2010/12/20
11	1925-M	19	39	2010/12/21
13	1937.5-N	43	35	2010/12/22
15	1962.5-S	85	51	2010/12/23
17	2025-M	52	43	2010/12/25
19	1925-M	43	48	2010/12/25
21	2012.5-N	108	58	2010/12/26
23	2012.5-N	45	47	2010/12/27
27	2212.5-M	45	39	2010/12/29
29	1962.5-N	46	36	2010/12/29
30	1950-M	78	45	2010/12/29

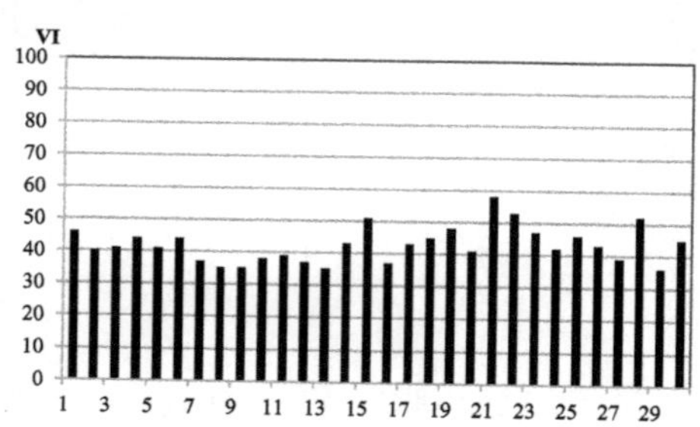

Figure 6. VI for 30 blasts, Sungun copper mine.

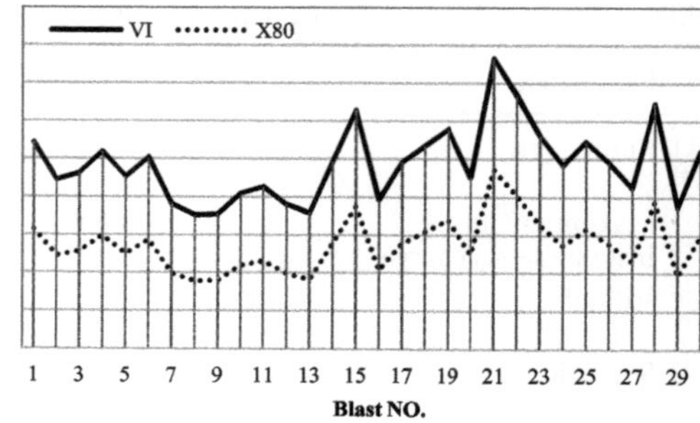

Figure 7. Agreement between the measured fragment size and VI for different blasts, Sungun copper mine.

blast No. 21 with maximum risk of 58 has the maximum X_{80} (108 cm), showing the capability of the newly model proposed.

5 CONCLUSIONS

The main conclusions of the study carried out are as follows:

1. The newly model presented in this paper, is an expert based model, which can deal with the inherent uncertainties in the geological systems. Also, it has the privilege of considering unlimited input parameters, which may affect on the system.
2. Applying RES in the rock blasting fields is a new and flexible method to find out the role of different numerous blast design parameters in a unit system.
3. Based on the model proposed, the calculated maximum and minimum vulnerability indexes for 30 blasts carried out in Sungun copper mine are 58 (for $X_{80} = 108$ cm) and 38 for ($X_{80} = 10$ cm) respectively. Comparison between these values confirms model efficiency in assessment of risk of fragmentation in blasting.

REFERENCES

Benardos, A.G. & Kaliampakos, D.C. 2004. A methodology for assessing geotechnical hazards for TBM tunneling-illustrated by the Athens Metro, Greece. International Journal of Rock Mechanics & Mining Sciences 41, 987–999.

Bhandari, S. 1997. Engineering rock blasting operations. A.A. Balkema, Rotterdam.

Cancelli, I. & Crosta, G. 1993. Hazard and risk assessment in rockfall prone areas. in: Ed.: Skipp, B., Risk and reliability in ground engineering, Thomas Telford, London, pp. 177–190.

Ghosh, A., Daemen, J.J.K. & Vanzyl, D. 1990. Fractal based approach to determine the effect of discontinuities on blast fragmentation. Proceedings of the 31st U.S. Symposium on Rock Mechanics, Golden, pp. 905–912.

Hudson, J.A. 1992. Rock engineering systems: theory and practice. Ellis Horwood, Chichester.

Hustrulid, W. 1999. Blasting principles for open-pit mining. A.A. Balkema, Rotterdam.

Jimeno, C.L., Jimeno, E.L. & Carcedo, F.J.A. 1995. Drilling and blasting of rocks. A.A. Balkema, Rotterdam.

Kulatilake, P.H.S.W., Qiong, W., Hudaverdi, T. & Kuzu, C. 2010. Mean particle size prediction in rock blast fragmentation using neural networks. Engineering Geology 114, 298–311.

Langefors, U. & Kihlstrom, B. 1978. The modern technique of rock blasting. John Willey & Sons, New York.

Latham, J.P. & Lu, P. 1999. Development of an assessment system for the blastability of rock masses. International Journal of Rock Mechanics and Mining Sciences 36, 41–55.

Latham, J.P., Munjiza, A. & Lu, P. 1999. Components in an understanding of rock blasting. Proceedings of the 6th International Symposium on Rock Fragmentation by Blasting, Johannesburg, South Africa, pp. 173–182.

Lu, P. & Hudson, J.A. 1993. A fuzzy evaluation approach to the stability of underground excavations. ISRM Symposium: EUROCK'93, Ed.: Ribeiro, L., Sousa, E., Grossmann, N.F., A.A. Balkema, Rotterdam, pp. 615–22.

Lu, P. & Latham, J.P. 1994. A continuous quantitative coding approach to the interaction matrix in rock engineering systems based on grey systems approaches. Proceedings of 7th International Congress of IAEG, A.A. Balkema, Rotterdam, pp. 4761–4770.

Mackenzie, A.S. 1967. Optimum blasting. Proceedings of the 28th Annual Minnesota Mining Symposium, Duluth, MN, pp. 181–188.

Mazzoccola, D.F. & Hudson J.A. 1996. A comprehensive method of rock mass characterization for indicating natural slope instability. QJEG 29, 37–56.

Sanchidrian, J.A., Seggara, P. & Lopez, M.L. 2007. Energy components in rock blasting, International Journal of Rock Mechanics and Mining Sciences 44, 130–147.

Shin, H.S., Kwon, Y.C., Jung, Y.S., Bae, G.J. & Kim, Y.G. 2009. Methodology for quantitative hazard assessment for tunnel collapses based on case histories in Korea. International Journal of Rock Mechanics and Mining Sciences 46, 1072–1087.

Singh, D.P. & Sastry, V.R. 1986. Influence of structural discontinuity on rock fragmentation by blasting. Proceedings of the 6th International Symposium on Intense Dynamic Loading and its Effects, Beijing, China.

Measurement and Analysis of Blast Fragmentation – Sanchidrián & Singh (Eds)
© 2013 Taylor & Francis Group, London, ISBN 978-0-415-62140-3

Optimisation of blast design for an iron ore mine and assessment of fragmentation through image processing

H.S. Venkatesh, K. Vamshidhar, G. Gopinath, A.I. Theresraj & R. Balachander
National Institute of Rock Mechanics, Bangalore, Karnataka, India

ABSTRACT: A non-captive iron ore mine wanted to increase their iron ore production by increasing their hole diameter from 102 to 165 mm. However, they were apprehensive about the deterioration in the input fragment size due to the shift to larger hole diameter. Field investigations were carried out for various blast designs practiced by the mine management and fragmentation was assessed using WipFrag a commercially available image processing software and by physical boulder counting method. Fragmentation analysis showed that the oversize fragments (>800 mm the input feed size of the crusher) percentage was about 5.4 and 19.7 for 102 mm and 165 mm diameter holes respectively. Subsequently, blast design modifications were suggested for 165 mm diameter holes and the fragmentation assessment showed that the oversize fragments were about 9.3% and hence the suggested modifications were beneficial to the mine. The paper deals with the modifications suggested and the assessment of fragmentation.

1 INTRODUCTION

Infrastructural growth of a country is primarily linked with steel, cement and power. With the accelerated development in India and neighbouring countries the demand for the iron ore too increased exponentially. One of the non-captive iron ore mines at Joda/Babri sector in Keonjhar district of Orissa is having a production capacity of 3000 tonnes per day at Jilling mine and 10,000 tonnes per day at Langalota mine. Jilling Langalota iron ore deposit comprises six ore bodies out of these three are of economic significance. These are—Langalota ore body, Gangaigora ore body and Appahatu ore body. The Langalota ore body is further divided into two sections, i.e. Langalota and Jajang. Besides these two and above three ore bodies, there are three more minor ore bodies namely—Jilling (almost exhausted), Rakhaburu and Khuntpani. This study relates to Langalota mine. The strata is very hard and massive on the western side benches while the eastern benches are highly fractured. A number of crushers are in operation to cater to the production needs. The blasted material is segregated, transported through dumpers and fed to different crushers according to their sizes. The input feed to the primary crushers is 800 mm. As the mine management was planning to enhance their production from the mine, they wanted to increase their drill hole diameter from the existing 102 mm to 165 mm. However, they were apprehensive about the deterioration in the input fragment size due to the shift to larger hole diameter.

2 ASSESSMENT OF FRAGMENTATION THROUGH IMAGE PROCESSING

Though physical counting of fragments can correlate very closely to the actual size distribution in the muckpile it is very cumbersome, time consuming and exhaustive. Considering these researchers have been working over a period of time to evolve methods to assess fragmentation. Image processing techniques proved to be suitable and effective. National Institute of Rock Mechanics (NIRM) has procured the latest version of WipFrag, which is proved to be commercially available across and has been using it in assessing the fragmentation at different sites. WipFrag accepts images from a variety of sources such as camcorders, fixed cameras, photographs, or digital files. It uses automatic algorithms to identify individual blocks, and create an outline 'net'. WipFrag allows manual editing to insert missing boundaries between fused fragments, and to delete false edges where a fragment has disintegrated into two or more pieces. It measures the 2-D net and reconstructs a 3-D distribution using principles of geometric probability. Figure 1 shows the generated net—edger detection and Figure 2 shows manually edited edge detection which is show as the identified fragments through Wipfrag.

The system allows various types of output according to individual requirements, Nominal diameter Dn, Mean fragment size D_{av}, Uniformity Index n, Characteristic size X_c, No of blocks, Maximum fragment size, Standard deviation of fragment size, Mode most common size particle,

Figure 1. Generated net—edge detection through WipFrag software for blast #1.

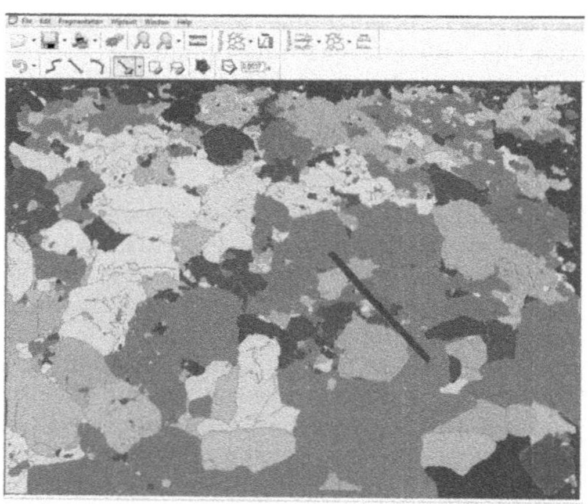

Figure 2. Identification of blocks and measuring the fragments through WipFrag software for blast #1.

including cumulative size distribution graphs and percentage passing at different sieve sizes (Wipware 1999).

3 MINE DETAILS

3.1 *Geology*

The stratigraphic sequence of the area is Banded Hematite Jasper (BHJ) with iron ore, purple shale, lavas and tuffs with dolerite dykes and unconformity. The area is a part of the westerly dipping eastern limb of the NNE plunging asymmetric synclinorium. The structure of the rock of this area is complicated due to several fold movements. The axis of the major fold is N-S. The ore bodies are laid down in the synformal with their axes plunging due south and north. Specifically the northern part of Langalota ore body is plunging due north and southern part is plunging due south. The ore bodies exhibit swelling and pitching structure with a maximum width of 600 m. The area shows extensive lateritisation at the top. The bedrock is concealed by laterite, shale and soil. However, some slickensides in the shales have been noticed indicating presence of minor strike faults. A thrust between volcanic formation and lower shale/chert of mixed faces formation is inferred from stratigraphic evidences. There are five sets of joints of which three sets are near vertical and two sets are near horizontal (Fig. 3). They are discernible in quarry faces in BHJ and massive iron ore. Big blocks of massive iron ore are seen standing as vertical columns showing columnar structure. The ore types are classified as Type 1 comprising of massive ore/hard laminated ore, Type 2 comprising of laminated ore/medium hard laminated ore, Type 3 comprising of soft and porous laminated ore. Type 4 is silliceous blue dust while Type 5 is BHJ/BHQ and Type 6 is shale. Physico-mechanical properties and Litho stratigraphy are given in Tables 1 & 2.

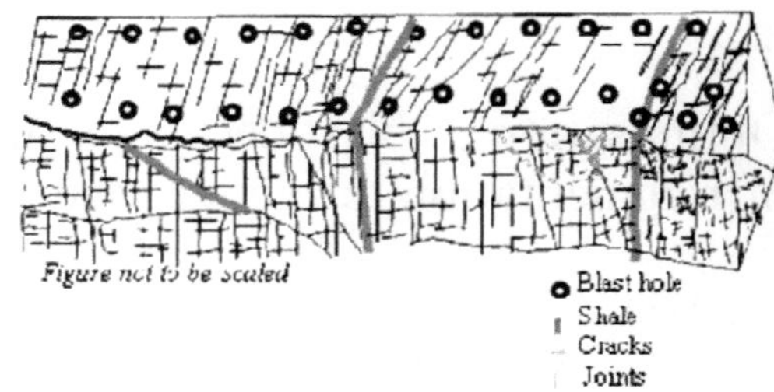

Figure 3. Block of blast 1 showing joints and bore holes.

3.2 *Method of working*

The mine is operated as a mechanised opencast using 102 mm diameter holes in 9 m benches. These holes are drilled using ROC F-9 Hydraulic Drill (Atlas Copco) or Pentra 1100 Hydraulic Drill (Sandvik). The drilled holes were charged and blasted with Site Mixed Emulsion (SME) and ANFO. Cartridged explosives were also used in some of the fractured holes. The blasted material is segregated, transported through dumpers and fed to different crushers according to their sizes. The oversize boulders were fragmented by secondary blasting by deploying Commando—120 H Drill machine. In some cases rock breaker was used as an alternative to secondary blasting.

Table 1. Physico-mechanical properties of rocks at Jilling Langalota Iron ore mine.

	Lithology			
Rock properties	Massive (Type 1)	Laminated ore (Type 2)	Soft & porous (Type 3)	BHJ (Type5)
Density, kg/m^3	5024	4999	4324	3536
Young's modulus, GPa	19	17	8	–
Tensile strength, MPa	6 to 11	5 to 10	2 to 4	4 to 6
Compressive strength, MPa	110 to 118	116 to 170	38	260 to 400
Shear strength, MPa	4 to 5	4 to 5	2.5	9

Table 2. Lithology of Jilling Langalota Iron ore mine.

	Thickness	
Litho unit formation	Average, m	Maximum, m
Shale (top) (Ore type 6)	12	20
Iron ore (Ore type 1, 2 & 3)	38	76
Shale parting (Ore type 6)	03	04
BHJ (Ore type 5)	02	03
Shale (Bottom) (Ore type 6)	04	07

3.3 *Review of blast design being practiced at the mine (102 mm drill hole diameter)*

Blast records for six months were collected from the mine. The data was categorized bench wise and details of the main parameters like burden, spacing and specific charge are given in Table 3.

From the data it may be observed that the bench height to burden ratio (>2 & <4) is acceptable. When the bench height to burden ratio is large, it is easy to displace and deform the rock, especially at the bench center. If H/B = 1, the fragments will be large, with overbreak and toe problems. With H/B = 2, these problems are attenuated and are completely eliminated when H/B >= 3 (Jimeno et al. 1995). The average specific charge used at the mine is below 1.0 kg/m^3. Drill cuttings are used as stemming material and stemming length was about 2 to 2.6 m. In many cases, diagonal initiation pattern was followed irrespective of orientation of joints.

3.4 *Fragmentation distribution baseline*

Though the above parameters play an important role in assessing the blast designs, the fragmentation size distributions can be considered as one of the most important key performance indicator. To establish the fragmentation distribution baseline three blasts were analyzed using WipFrag. Table 4 gives the details of the blast geometry in these blasts.

For each blast randomly twenty images were clicked while loading to dumper. Images which are bad quality poor lighting and uneven frames are manually rejected and minimum ten to fifteen images are analysed and finally the results of the individual frames were merged to a single frame to obtain the graphs of average size distribution of the blast. In order to digitally sieve the fragments it's essential to have a reference scale and we have used a reference scale 1 m × 1 m in dimension. The size distribution for each image is manually edited of delineation fragments after auto netting. The time spent for each image is about 30 minutes. Figure 4 shows the size distribution curve obtained through WipFrag software for three blasts. It may be noted that about 96.5% of the fragments are below 800 mm which happens to be the sieve size of the grizzle in the mine for blast 1. The same approach was followed for the other two blasts and the results are given in Table 5. The mine blasts shows an average oversize value (>800 mm) of 5.4% to be considered as the fragmentation baseline. The fragmentation obtained with the increased hole diameter (165 mm) for the blast designs from the mine authorities and for the modified designs from NIRM were measured against this baseline value of 5.4% of oversize material.

3.5 *Assessment of fragmentation for mine blasts—165 mm drill hole diameter*

The main purpose of this study was to ascertain whether the mine can induct 165 mm diameter blasthole drill and increase the production without compromising the fragmentation. Further the mine was constrained with regard to bench height as they were not in a position to increase it from the current 9 m height. The mine procured a 165 mm diameter blasthole drill (DI 600/C15-ACERT Sandvik drill machine) and designed the blasts (Table 6). These blasts were assessed for the fragmentation size distribution through image

Table 3. Summary of blast parameters (102 mm hole dia).

Parameters (Average)	Working benches RL 491	RL 500	RL 509	RL 518	RL 527	All benches
Bench height, m	7.0	6.9	7.9	8.3	7.0	7.0
Burden, m	2.6	2.6	2.7	2.7	2.7	2.7
Spacing, m	3.1	3.1	3.1	3.2	3.2	3.2
Burden to hole diameter ratio	25.5	25.5	26.5	26.5	26.5	26.5
Bench height to burden ratio	2.7	2.7	2.9	3.1	2.6	2.6
Specific charge, kg/m³	0.86	0.87	0.84	0.82	0.80	0.83

Table 4. Details of the blasts analysed (102 mm dia).

Parameters	Blast-1	Blast-2	Blast-3
Ore type*	Type 2 & 5	Type 1	Type 2
Hole depth, m	9.1	9.5	6.9
Bench height, m	8.5	8.5	6.0
Sub grade, m	0.6	1.0	0.9
Drilling pattern	Staggered	Staggered	Staggered
Total no of holes	28	31	43
Number of rows	3	3	3
Burden, m	2.5	2.4	2.5
Spacing, m	3.0	3.1	3.1
Stemming, m	2.6	2.0	1.9
Stemming material	Drill cuttings	Drill cuttings	Drill cuttings
Charge per hole, kg	60.5	53.4	42.3
Charge length, m	6.5	7.5	5.0
Total charge, kg	1695	1655.6	1820.4
Volume, m³	1785	1960	2000
MCD♠, kg	242	213	169
Initiation sequence	Diagonal cut	Diagonal cut	Diagonal cut
Delay timing, ms	25	25	17 & 42
Burden to hole diameter ratio	24.5	23.5	24.5
Bench height to burden ratio	3.4	3.5	2.4
Specific charge, kg/m³	0.95	0.84	0.80
Major joints, cm	2 to 15	2 to 15	2 to 5
Vertical mild joints, mm	2	2	2
Shale band, cm	2 to 40	2 to 40	2 to 25

Ore type*: Ore types are given in section 2.1, MCD♠: Maximum charge per delay.

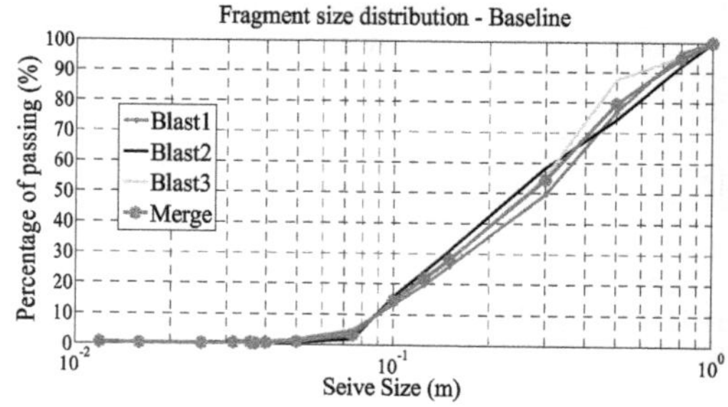

Figure 4. Average Fragment size distribution for 102 mm drill hole diameter for different blasts.

Table 5. Summary of fragment size distribution for 102 mm drill hole diameter.

Fragment Size, mm	Passing percentage of fragments Blast-1	Blast-2	Blast-3	Average
1000	99.9	100	100	100
800	**96.5**	**92.1**	**95.2**	**94.6**
500	77.0	74.4	87.5	79.6
300	49.5	58.4	55.4	54.4
150	25.8	31.0	27.6	28.1
125	19.8	23.9	21.8	21.8
100	13.3	15.7	14.7	14.6
X_{50}	0.310	0.275	0.239	0.274
No of blocks	962	1034	1044	1013
n	1.82	1.84	2.19	1.95

processing. The results of the individual frames were merged to get combined result for obtaining the average size distribution in each blast. The details of analysis with regard to fragmentation are given in Table 7. It can be seen that the mean fragment size is less than 500 mm (200 mm more than that with 102 mm diameter holes) and the average percentage passing (80.3%) is lower

Table 6. Drilling and blasting details followed for 165 mm drill hole diameter (Mine design).

Parameters	Blast-1	Blast-2	Blast-3	Blast-4
Ore type*	Type 2 & 6	Type 1, 2, 3, 5, 6	Type 1, 3, 5, 6	Type 1 & 6
Hole depth, m	9.8	9.1	7.7	10.0
Bench height, m	9.0	8.0	7.0	9.0
Sub grade, m	0.8	1.1	0.7	1.0
Drilling pattern	Staggered	Staggered	Staggered	Staggered
Total no of holes	28	42	36	47
Number of rows	2	3	3	3
Burden, m	4.5	4.5	4	3.9
Spacing, m	5.0	5.5	4.5	5.2
Stemming♠, m	4.5	3.4	3.3	3.3
Charge per hole, kg	164.6	16.0	100.8	145.6
Total charge, kg	4607	6760.5	3629.5	6844
MCD♠, kg	658	483	302	582
Initiation sequence	V cut	Diagonal cut	Diagonal cut	Diagonal cut
Delay timing, ms	17	25	25	25
Burden to hole diameter ratio	27.3	27.3	24.2	23.6
Bench height to burden ratio	2.0	1.8	1.7	2.3
Volume, m^3	5670	8316	4536	8578
Specific charge, kg/m^3	0.81	0.81	0.80	0.80
Major joints, cm	2 to 3	2 to 3	2 to 3	2 to 3
Vertical mild joints, mm	2	2	2	2
Shale band, cm	5 to 15	5 to 25	2 to 5	5 to 15

Ore type*: Ore types are given in section 2.1, Stemming♠: Drill cuttings, MCD♠: Maximum charge per delay.

Table 7. Summary of fragment size distribution (Mine blasts 165 mm hole diameter).

Sieve	Passing percentage of fragments				
Size, mm	Blast-1	Blast-2	Blast-3	Blast-4	Average
1000	81.0	89.9	88.2	96.8	88.7
800	**75.6**	**83.2**	**80.0**	**82.5**	**80.3**
500	60.3	58.6	57.9	58.4	60.1
300	26.9	30.8	31.9	30.6	35.5
150	11.7	12.7	10.5	12.4	12.9
125	07.8	09.9	05.8	09.6	08.7
100	03.8	06.4	02.6	06.1	04.7
X_{50}	0.389	0.40	0.42	0.425	0.408
No of blocks	1711	1767	557	1130	1291
n	1.34	1.44	1.97	1.34	1.52

than the baseline (94.6%). In other words with the induction of 165 mm diameter drill machine the percentage oversize fragments increased from 5.4 to 19.7. The mine management was worried as their apprehensions with regard to deterioration in fragmentation due to increased hole diameter came true. Subsequent to the trials of the mine authorities with 165 mm diameter blastholes, NIRM suggested suitable designs. The details of the trial with the modified designs are discussed in section 3.6 (Venkatesh et al. 2009).

3.6 *Assessment of fragmentation for modified blasts—165 mm hole*

NIRM modified the designs and carried out four blasts with 165 mm diameter blast holes. The burden and spacing were designed according to the ore type and depth of holes. The holes were charged with SME and cartridge explosives. As opined, angular gravel/aggregate of 10–20 mm size were used as stemming material. The average charge per hole varied from 148 to 206 kg, and the total charge in a round varied between 3925 to 9978 kg. The blasts were initiated with shock tube initiation system (Twin dets). Due to fractured strata, cartridge explosives were also used to build up the column. The details of the blasts are given in Table 8.

Table 8. Blast design parameters followed during the trial blasts (Modified design—165 mm drill hole diameter).

Parameters	Blast-1	Blast-2	Blast-3	Blast-4
Ore type*	Type 2	Type 2, 5	Type 2, 6	Type 1, 6
Hole depth, m	8.6	7.5	9	11
Bench height, m	7.3	6.3	8	11
Sub grade, m	1.3	1.2	1	0
Drilling pattern	Staggered	Staggered	Staggered	Rectangle
Total no of holes	44	35	63	19
Number of rows	3	3	2	2
Burden, m	4.1	3.9	3.6	3.5
Spacing, m	5.0	4.9	4.7	4.3
Stemming♠, m	3.0	2.7	2.9	2.6
MCD♠, kg	701	711	1205	878
Initiation sequence	V cut	Row by row	Row by row	V cut
Delay timing, ms	42	42	42	25
Burden to hole diameter ratio	24.8	23.6	21.8	21.2
Bench height to burden ratio	1.8	1.6	2.2	3.1
Specific charge, kg/m³	0.84	1.14	1.04	1.25
Major joints, cm	2 to 5	2 to 5	2 to 5	2 to 5
Vertical mild joints, mm	2	2	2	2

Ore type*: Ore types are given in section 2.1, Stemming♠: angular gravel/aggregate of 10–20 mm size, MCD♠: Maximum charge per delay.

3.6.1 *Trial blast 1*

The length and the width of the block was 90 m and 12 m. Blast was initiated with 'V' pattern with 17 ms and 25 ms delay within the row and 42 ms between the rows as shown in Figure 5. From the video observation (Fig. 6) of the blast it can be observed that the block moved well. Stemming ejection was noticed in the middle and towards south side of the blast where the bench height was low. The designed specific charge appeared to be normal. The fragmentation was excessive in the middle portion of the block indicating scope for reducing the specific charge in the subsequent blasts. Muckpile was good and the throw was observed to be about 30 m which was well within the bench limit.

As the block was initiated with 'V' pattern it was observed during mucking that the fragments were finer, uniform and compacted due to collision. Probably, row by row initiation might have resulted in more uniformly distributed coarser fragments with a flatter muckpile. Straight and neat face was obtained after mucking. Back break was not observed. Charge factor was 16.2% in excess of the designed due to fractured strata and also due to enlargement of blastholes.

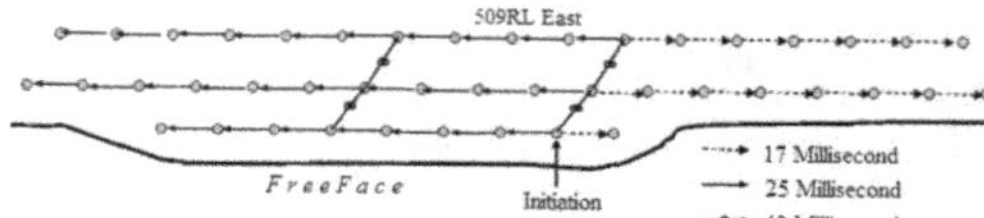

Figure 5. Drilling, hookup pattern for the blast #1.

Figure 6. Video observation for blast #1.

3.6.2 *Trial blast 2*

The length and the width of the block was 65 m and 12 m with three rows drilled in staggered pattern. The block was initiated row by row with four holes together and 17 ms and 25 ms delays were used within the rows while 42 ms delay was used between the rows (Fig. 7). One hole in the middle portion of the front row was re-drilled as it was choked. Unfortunately while charging, the stemming rod fell into the hole resulting in poor explosive column build up. This resulted in tight muck around this hole. Similarly, one hole at the

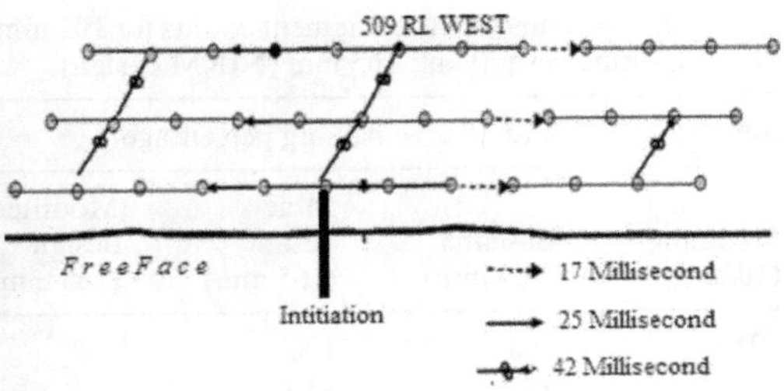

Figure 7. Drilling, hookup pattern for the blast #2.

Figure 8. Video observation for blast #2.

last row could not be charged fully as the hole was choked and re-drilling was also not possible due to site conditions. From the video observation (Fig. 8) it can be seen that that stemming ejection was predominant due to fractured nature of the strata and the use of drill cuttings. Back break was observed at the locations where the holes had collapsed in the front row. Fragmentation was good and some boulders were observed without fresh broken surfaces indicating embedded boulders in soft ore. In practice a 165 mm diameter hole can accommodate about 22 kg/m at a density of 1.04 g/cc. However in this case a charge of 33.7 kg/m was accommodated due to cavity and hole enlargement. Hence this increased the explosive per meter in the column by about 52% leading to the increase in specific charge to about 1.13 kg/m^3. This specific charge happens to be on the higher side and this could be reduced by using doped Emulsion or cartridge explosive or ANFO/SME loaded in a 150 mm polythene sleeve. Muck profile was good except in the middle portion and the throw was observed to be about 20 m.

3.6.3 *Trial blast 3*

The length and width of the block was 125 m and 7 m with two rows drilled in staggered pattern. Video observation reveals that the stemming ejection was noticed due to variation in stemming length range from 2.0–3.5 m (Fig. 9). The block

Figure 9. Video observation for blast #3.

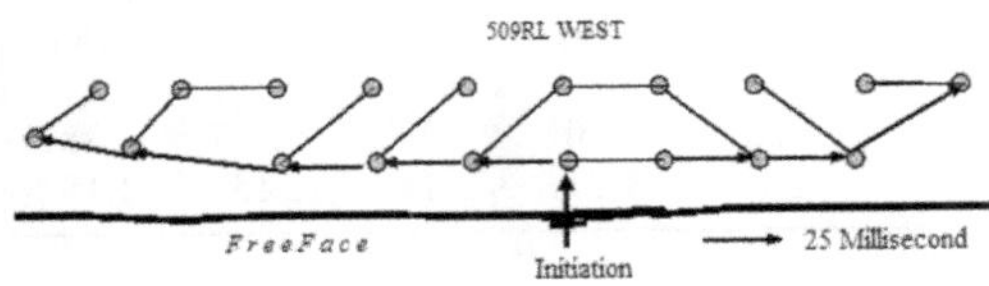

Figure 10. Drilling, hookup pattern for the blast #3.

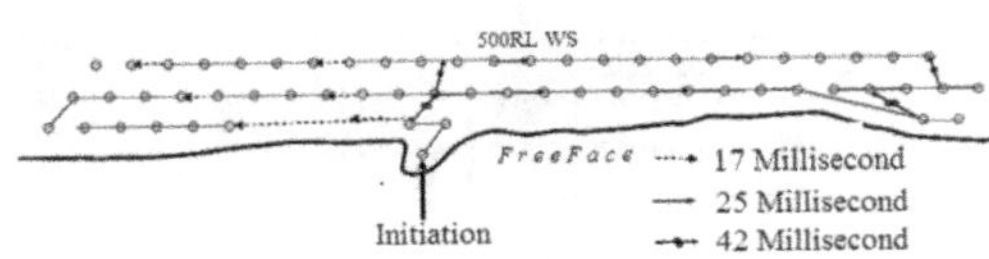

Figure 11. Drilling, hookup pattern for the blast #4.

was initiated with five holes per delay. Sequence of initiation is shown in Figure 10. The explosive was about 18% in excess than the designed quantity due to fractured strata and enlargement of blastholes. No back break was observed. Muck profile was good, and the throw was observed to about 30 m.

3.6.4 *Trial blast 4*

The length of block was 50 m while the width was only 7 m. Considering this width only two rows could be drilled. Holes were drilled in a rectangular pattern and initiated with V1 sequence (Fig. 11) keeping in view the hard strata (massive with widely spaced joints). It was observed that about 12% of explosives was charged excessively than the designed quantity due to fractured strata and also due to enlargement of hole diameter. Muck profile was good, no back break and the throw was observed to about 30 m. Over all the blast was good with good fragmentation.

3.7 *Comparison of fragment size distribution with modified design for 165 mm hole diameter*

The passing percentage of fragments with the modified design for 165 mm is given in Table 9. From Figure 12, it can be observed that the

Table 9. Summary of fragment size distribution for modified design—165 mm drill hole diameter.

Sieve	Passing percentage of fragments				
Size, mm	Blast-1	Blast-2	Blast-3	Blast-4	Average
1000	100	92.2	95.9	96.9	96.2
800	**97.3**	**87.3**	**90.0**	**88.2**	**90.7**
500	85.4	78.2	74.6	81.2	79.9
300	59.4	53.9	49.6	58.3	35.3
150	15.3	13.5	10.6	17.9	14.3
125	08.2	08.0	05.1	10.4	07.9
100	04.6	04.1	02.1	04.7	03.9
No of blocks	1555	1499	1559	1702	1785
X_{50}	0.260	0.307	0.302	0.257	0.281
n	1.11	1.21	1.14	1.15	1.15

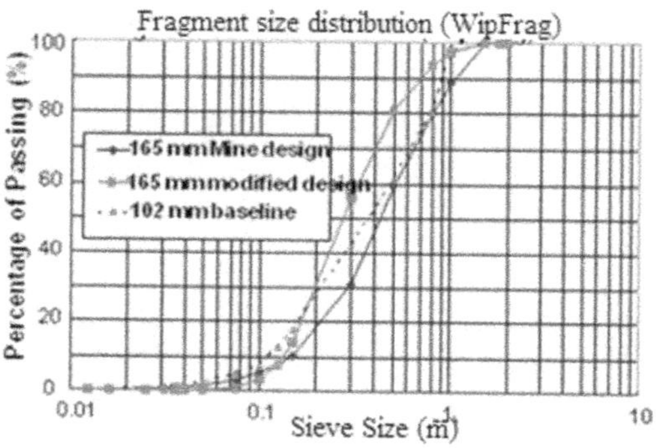

Figure 12. Comparison of fragment size distribution for 102 mm (baseline), 165 mm (Mine design) and modified design.

fragmentation has improved considerably with the modified design. Comparing the mine design and the modified design by the authors, it can be seen that there has been an increase in passing percentage by 10.4 (90.7–80.3) for the given grizzle size of (800 mm). On comparing the percentage passing of the fragments from the modified design with the baseline fragmentation percentage passing, the difference is 3.9 (94.6–90.7). This shows that even with the increased blasthole diameter of 165 mm the fragmentation sizes as compared to 102 mm diameter are similar and hence the apprehension of the mine owner is allayed.

3.8 *Random physical count of fragments*

As a cross check to the assessment by image analysis, random physical counting of fragments for each blast was carried out to have a feel of the blast in terms of fragmentation size distribution.

Table 10. Summary of the fragment counts for 102 mm, 165 mm (mine design) and 165 mm (NIRM design).

Sieve	Cumulative passing percentage		
Size, mm (102 mm)	Baseline (165 mm)	Mine design (165 mm)	Modified design (165 mm)
1259	100	100	100
1144	93.7	93.8	98.2
1000	89.9	91.5	95.3
928	87.0	88.0	93.5
800	79.0	80.0	90.0
669	62.8	62.5	79.2
464	32.4	25.0	43.0

The procedure followed was to divide the muck pile into three sections namely top, middle and toe. In each section, different fragments were randomly counted such that it covers complete exposed blasted materials. For each blast, about 100 fragments were measured for their length, width and thickness. The analysis of the measured fragments from all the blast is given in Table 10. It can be noted that the mine blasts shows an average oversize value (>800 mm) of 21% for 102 mm diameter blast holes. The fragmentation obtained with the increased hole diameter (165 mm) for the blast designs from the mine authorities is 20% where as the blasts designed by authors is only 10% of oversize material.

4 RECOMMENDED BLAST DESIGN PARAMETERS

The recommended blast design for the mine is given in Table 11. However in order to enhance the efficiency of the suggested design some of the suggestions are as below:

a. To increase the bench height from the current 9 to greater than 12 m.
b. It was observed in the field that 165 mm diameter holes were getting enlarged almost to about 200 mm diameter due to the fractured nature of the strata. This is invariably increasing the specific charge and hence it was suggested to use:
    i. Cartridge explosive to build up the explosive column.
    ii. Light density ANFO or doped emulsion.
    iii. Smaller drill bit so as to get a finished diameter of 165 mm.
    iv. Polythene sleeves to avoid explosives getting wasted through voids and cracks.
    v. Casing the disturbed holes with PVC pipes.

Table 11. Recommended blast design parameters for different ore types.

Parameters	Hard ore with vertical joints	Medium ore	Highly fractured ore	Embedded boulders
Bench height, m	9.0	9.0	9.0	9.0
Hole depth, m	10.0	10.0	10.0	10.0
Sub grade, m	1.0	1.0	1.0	1.0
Drilling pattern	Staggered	Staggered	Staggered	Staggered
Burden, m	3.5	3.5	4.0	3.5
Spacing, m	4.5	4.5	5.0	3.5
Drill factor, m^3/m	15.7	15.7	20.0	12.2
Booster charge per hole, kg	0.25	0.25	0.25	0.25
Column charge (SME) per hole, kg	153	143	143	117
Charge length, m	7.0	6.5	6.5	6.5
Stemming*, m	3.0	3.5	3.5	3.5
Initiation sequence	V	Row by row	Row by row	Row by row
Delay timing, ms within rows	25	25	25	25
Between rows, ms	42	42	42	42
Specific charge, kg/m^3	1.08	1.01	0.79	1.06

Stemming*: angular gravel/aggregate of 10–20 mm size.

5 CONCLUSIONS

The recommended blast designs have ensured the desired fragmentation even with the increased blasthole diameter (102 mm to 165 mm). Fragmentation assessment through image processing showed that the passing percentages at 0.8 m sieve size with the modified blast design was 90.7% comparable with that of the baseline passing percentage of 94.6. Considering the burden stiffness due to shallow bench height, it is appropriate to increase the bench height to greater than 12 m. The problem of increased blasthole diameter due to strata condition can be tackled by use of light density ANFO or doped emulsion.

ACKNOWLEDGEMENTS

The authors are thankful to Essel Mining & Industries Ltd. for providing necessary facilities during filed investigation. The permission of Director, National Institute of Rock Mechanics, India to publish this paper is acknowledged.

REFERENCES

Jimeno, C.L. Jimeno, E.L. & Carcedo, F.J.A. 1995. *Drilling and blasting of rocks*. Rotterdam: Balkema.

Venkatesh, H.S., Theresraj, A.I., Gopinath, G., Vamshidhar, K. & Balachander, R. 2009. Final Report on Optimisation of blast design for Jilling Iron and Manganese mines, Essel Mining and Industries Ltd, Keonjhar, Orissa. Project No RB 0806C (Unpublished), September.

Wipware. 1999. WipFrag & WipJoint operation Manual.

Measurement and Analysis of Blast Fragmentation – Sanchidrián & Singh (Eds)
 ISBN 978-0-415-62140-3

Research on prediction of average blasting fragmentation based on BP neural network

G.Q. Zhang & T.J. Tao
University of Science & Technology Beijing, Beijing, China

X.G. Wang & C.P. Wu
Beijing General Research Institute of Mining & Metallurgy, Beijing, China

ABSTRACT: A Back-Propagation (BP) neural network is applied to the prediction of average blasting fragmentation. Burden, powder factor and maximum charge per hole are considered as the main factors to establish the back-propagation neural network model in the analysis. The BP neural network that has been established previously is trained, taking the deep-hole blasting in an open-pit mine as experimental object by using the raw information and the monitoring data of average blasting fragmentation collected in the detonation process. Finally, the trained model is applied to predict the blasting fragmentation. Results show that the forecast data by the BP neural network model are very close to the actual values and that it is an effective way to predict the average blasting fragmentation that can fulfill the request of engineering practice preferably.

1 INTRODUCTION

Blasting fragmentation is a main index in estimating blasting quality. The prediction and controlling of fragment size have become one of the main research subject to acquire optimal blasting effect. The requirement of fragment size will change according to different engineering object, such as convenience in shipment loading, crushing and so on (Xie & Cai 1995). Scholars home and abroad have engaged in abundant research about blasting fragmentation. Distribution function of fragment size, slow motion on blasting, statistics of block rate, fuzzy mathematics, grey system theory et al were adopted to ascertain fragment size distribution and evaluate blasting effect.

Zhou (2003) presented a method of predicting the blasting fragmentation based on the simulation test of small field. Through regression analysis, a predicting equation of blasting fragmentation is established between blasting fragmentation distribution and seven variables including the rock cracks, type of explosive, borehole spacing, subdrill, etc. At the same time, according to the predicting equation, the calculation formulae of block rate, average size of fragments and P_{80}, k_{50} were obtained.

Taking into account the jointed rock mass structure and local nonlinear dynamic interaction of rock Guo et al. (1997) built a mechanical model of fragmentation forecast on jointed rock mass by blasting with the multiple rigid dynamic and Bond's fragmentation theory. Firmly combining to jointed characteristic, blasting mechanism with fragmentation calculation, the insufficiency of their mutual disjoint previously is avoided.

Fully considering the breaking action of the explosion stress wave and the moving fragmented blocks extrusion and collision under the action of the detonation gas expansion and infiltration flow pressure, Zhang & Chen (2002) proposed a theoretical calculation model of the fragmentation size distribution containing damage variable. The average fragment size (k_{50}) and the percentage of undersize are derived from the dynamic average fragmentation size formula for fragile materials given by Grady on the basis of fragmentation energy rules. Some numerical simulations and verification tests are carried out. The results of numerical simulations coincide well with testing data. Simulations and tests show that the action of the extrusion and collision between moving fragment blocks is a significant factor affecting the average fragmentation size regardless the type of rock mass, especially for the rock mass with very bad quality.

Through discussing the fractal trend of fragments and the fractal characteristics of rock blasting fragmentation, Xie & Cai (1995) found that the fractal geometry might provide a new powerful approach for forecasting rock fragmentation by blasting. And then they applied fractal method to study the relation between blasting efficiency and blasting fragmentation.

However, it is very difficult to predict fragment size, because there are many correlating factors to control blasting fragment size, which degree of influence is not well known. Although much work has been done, the results are less than satisfactory, so further analysis and study should be conducted (Zhang 1992a). In this paper, BP neural network was introduced to predict average fragment size, in an effort to advance the prediction accuracy and simplify the forecasting method.

2 BP NEURAL NETWORK

BP neural network was put forward by a select group of scientists headed by Rumelhart and McClelland in 1986 (Rumelhart et al. 1986b); it consists of a multilayer feed-forward network trained as error back-propagation algorithm, and one of the most popular neural network. BP neural network can study and store plentiful mapping relationship of input and output model without describing the equation of the mapping relationship. It completes system prediction through searching for quantity expression of system input and output according to network trained data, with self-adaptive and study process.

BP network consists of nonlinear transformation units. It is based on a gradient descent method with self-learning algorithm. There are input layer nodes, output layer nodes and one or more hidden layer nodes. Input information is disposed in hidden layer, and then passed to output layer. Each layer of neural just influences the next layer of the neural (Rumelhart et al. 1986a).

If a desired output can't be obtained, it turns to back propagation, which makes the error information minimum through modifying weight of each layer of neural. Supposing there are P network input samples, X^k is the input information vector, T^k ($k = 1,2, ..., P$) is the expected output information vector, then the algorithm steps of BP neural network can be described as follows (Duan et al. 2010):.

1. Initiation of network, initial value of connection weight and threshold are assigned randomly.
2. Input researching samples, compute input u_j and output o_j of each layer:

$$u_j^k = \sum_i W_{ij} o_i^k - \theta_j \tag{1}$$

$$o_j^k = f(u_j^k) = f\left(\sum_i W_{ij} o_i^k - \theta_j\right) \tag{2}$$

where W_{ij} is the connection weight of nodes i and j, i the upper layer node, j the lower layer next to node i; θ_j the threshold value of node j and f the action function of node.

3. Compute the total error of the network E by:

$$E = \sum_{k=1}^{p} E_k = \frac{1}{2}\sum_{k=1}^{p}\sum_{t=1}^{n}\left(t_i^k - y_i^k\right)^2 \tag{3}$$

If the total error $E \geq e$ or single error $|t_i^k - y_i^k| \geq e$ go to step 4; else the demand can be fulfilled, and network exports.

4. Compute the gradient modification error of each layer according to Equations 4 and 5, and then modify the weight.

$$\frac{\partial E_k}{\partial W_{ij}} = \frac{\partial Ek}{\partial u_j^k}\frac{\partial u_j^k}{\partial W_{ij}^k} \tag{4}$$

$$W_{ij}^{(c+1)} = W_{ij}^{(c)} + \eta\frac{\partial E}{\partial W_{ij}} = W_{ij}^{(c)} + \eta\sum_{k=1}^{p}\delta_j^k o_i^k \tag{5}$$

where c is iteration coefficient, η learning step and $\delta_j^k = \partial E_k/\partial u_j^k$ gradient modification error.

3 CONSTRUCTION OF THE NEURAL NETWORK

3.1 *Brief project overview*

Fangmayu iron ore is an open-pit mine belonging to Miyun Mine & Metallurgy Corporation, and is located on the north of Gaolin town, Miyun county. Geographic coordinates of the ore are: East longitude 117°07′30″ to 117°11′15″, Northern latitude 40°25′00″ to 40°27′20″. Mining area is about 4.7 km^2. Low hills are the main topography in the mining area and the elevation is around 140 to 320 m with 20 to 100 m of relative relief. What's more, the ravine in the mining area is wide. Ground water empties into Miyun Reservoir through Chao River from North to South. Magnetite is the main mineral. Joint fissures are well developed inside the ore and overburden. Borehole diameter is 90 mm. Mining area environment is shown in Figure 1.

3.2 *Determination of nodes in input and output layer*

Self-characteristics of rock mass affect the average blasting fragmentation directly, such as jointing, fracture and so on, which are difficult to acquired and quantize. The affection of rock mass characteristics can be eliminated through enhancing training of model by BP neural network.

Figure 1. Mining area environment.

In other words, target information can be obtained after some scale of network training is conducted, so self-characteristics of rock mass can be ignored when constructing BP neural network. However, if rock mass characteristics change, the established BP neural network should be trained again in order to predict the average blasting fragmentation accurately.

Some typical and main influence factors can be considered to established BP neural network model combining with engineering practice experience. In the model, line of least resistance (burden) W, unit explosive consumption (powder factor) q and maximum charge per single hole Q are selected as nodes of input layer through analyzing the main factors that affect the average fragment.

Because average fragment size is the only research target in this paper, there is only one node in output layer. The data of average fragmentation is derived with fragment size tester researched and developed by Guangdong Honda Blasting Co., Ltd. The photo of real object is shown in Figure 2 and the operation interface diagram of the tester is shown in Figure 3.

3.3 *Determination of hidden layer nodes*

A BP neural network with one hidden layer can achieve mapping arbitrarily from one-dimensional to m-dimensional in some extent (Zhang 1992b), so a BP network with one hidden layer was selected when constructing the average fragment forecasting model. There is no computational formula in theory. Generally speaking, the more the hidden layer nodes are, the better the prediction result, but excessive nodes will increase calculated amount so as to influence the convergence rate of network. On the contrary, prediction error is likely to be magnified because of the reduction of generalization ability along with

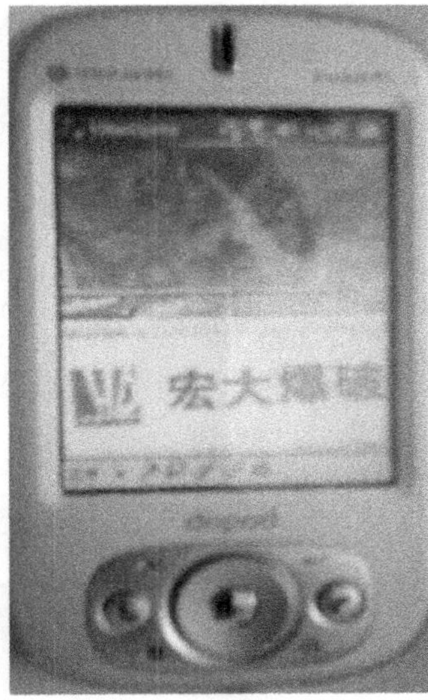

Figure 2. Blast sprite mobile PDA.

Figure 3. Operation interface diagram of the tester.

overfitting. Consequently, the number of hidden nodes should be reduced as little as possible on the promise that the predictive validity would be guaranteed. The number is no less than that of output nodes, which can be extracted according to (Young et al. 2000):

$$n = 2m + 1 \tag{6}$$

where n is the number of hidden nodes and m the number of input nodes.

The number of hidden nodes is 7 according to Equation 6. Then the model consists of three layers, which includes three input nodes, one output nodes and seven hidden nodes to ensure the precision of network computing.

4 TRAINING AND APPLICATION OF THE BP NEURAL NETWORK MODEL

4.1 *Training of the network*

If only the samples be selected, training can be conducted according to model structure and algorithm after basic structure of predictive model was determined. The more the training samples, the more precise the predicted result. In this paper, the average fragment sizes of Fangmayu open-pit iron ore were obtained through photographic technique and image processing and analyzing, in which 22 groups of data were selected as training samples, shown in Table 1.

Matlab neural network toolbox was used to program and calculate in order to train and apply the model. Linear function was selected as the action function of output nodes, while S-mode nonlinear function was selected as the action function of hidden nodes. The input data and desired output were all normalized before calculation so as to reduce iteration times and guarantee

Table 1. Training samples of BP model.

Serial number	Burden (m)	Powder factor ($kg \cdot m^{-3}$)	Maximum charge per hole (kg)	Average fragment (cm)
1	2	0.329	12.0	46.7
2	2	0.361	13.5	44.6
3	1.5	0.496	12.5	33.8
4	2	0.476	12.0	35.0
5	2.5	0.343	12.0	44.9
6	1.8	0.345	13.2	45.6
7	2	0.315	10.0	46.9
8	2.1	0.379	6.0	37.4
9	2.3	0.306	9.0	47.5
10	2.2	0.429	13.0	38.7
11	1.9	0.383	11.2	41.2
12	2.1	0.463	12.6	36.1
13	1.6	0.498	12.5	34.0
14	2.1	0.365	11.0	42.3
15	1.9	0.392	11.6	40.5
16	2.3	0.337	10.7	45.1
17	2.5	0.309	12.1	50.1
18	2	0.372	10.3	41.5
19	2.3	0.325	13.2	48.2
20	2.1	0.320	10.9	47.1
21	2.3	0.339	9.0	43.7
22	2.4	0.306	13.0	50.2

Table 2. The forecast results of BP model.

Serial number	Burden (m)	Powder factor ($kg \cdot m^{-3}$)	Maximum charge per hole (kg)	Forecast value (cm)	Measured value (cm)	Relative error (%)
1	2.3	0.361	12.2	43.3	46.4	6.68
2	1.9	0.377	11.2	41.4	39.0	6.15
3	2	0.364	11.5	42.6	50.2	15.14
4	2.5	0.306	12.5	49.9	55.3	9.76
5	2.2	0.367	11.9	42.7	35.1	21.65
6	2.1	0.385	12	41.4	39.9	3.76
7	1.9	0.398	10.6	40.1	47.4	15.40
8	2.2	0.37	12.1	42.5	40.2	5.72
9	1.8	0.438	10.7	37.8	40.2	5.97
10	1.5	0.443	10.3	37.5	38.1	1.57
11	2.1	0.376	11.5	41.9	38.8	7.99
12	1.9	0.395	11.2	40.5	45.6	11.18

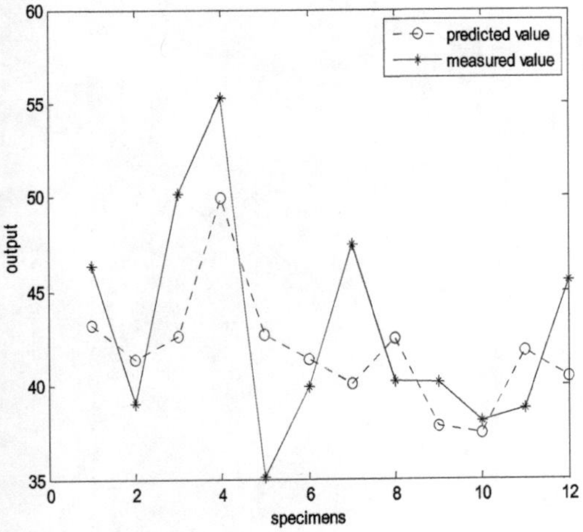

Figure 4. Comparison of forecast results and actual values.

convergence of algorithm because the value range of each input nodes variety in large scale. Accuracy of model was selected as 0.0001, and study steps 0.1. Training does not finish until precision requirements are met after several hundreds of iterations. Then connection weight of each node was ensured. The model can be used to predict similar samples.

4.2 *Application of the network*

12 groups of data similar to the training samples were selected randomly as input data that was imported into the trained model to predict average fragment size. Forecasting results and errors are listed in Table 2.

Results show that the relative error between predicted value of the BP network model and measured value is small. All the relative errors are less than 21.65% among the 12 selected samples. The predicted value of the BP network is very close to the measured value according to the samples serial curve (shown in Fig. 4), which is in keeping with the distributed trend of measured value.

5 CONCLUSION

The influence factors of blasting fragment size are very difficult to describe with simple theory model. However, the problem can be handled easily with the use of BP neural network, which demonstrate the advantage of neural network theory. Calculating and analyzing results show that blasting fragment size on all kinds of geological conditions can be predicted by BP neural network, which can fulfill the forecasting demands of routine production. The selection of influence factors is very flexible. If more factors that influence blasting fragment size were considered as input nodes of the BP neural network, predicted precision can be heightened. If more parameters such as block rate, etc. were considered as output in training samples, then such output results can also be obtained in the BP neural network prediction, which can provide more direction and reference for production blasting.

REFERENCES

Duan, B.F., Zhang, M. & Li, J.M. 2010. A BP neural network model for forecasting of vibration parameters from hole-by-hole detonation. *Explosion and Shock Waves* 30(4): 401–406.

Guo, W.Z., Wang, S.R. & Liu, D.S. 1997. Discussion on the mechanical model of jointed rock mass fragmentation forecast by blasting. *Blasting* 14(3): 31–34.

Rumelhart, D.E., Hinton, G.E. & Williams, R.J. 1986a. Learning representations by back-propagating errors. *Nature* 323: 533–536.

Rumelhart, D.E., McClelland, J.L. & the PDP Research Group. 1986b. Parallel Distributed Processing: Explorations in the Microstructure of Cognition. Volume 1: Foundations. Cambridge, MA: MIT Press.

Xie, X.P. & Cai, J.S. 1995. Application of fractal method on forecasting. *Journal of Tangshan Institute of technology* 12(1): 6–11.

Young, M.T., Blanchard, S.M., White, M.W., Johnson, E.E., Smith, W.M. & Ideker, R.E. 2000. Using an artificial neural network to detect activations during ventricular fibrillation. *Computers and Biomedical Research* 33(1): 43–58.

Zhang, J.C. 1992a. Summary of fragment-size predicting model in rock mass blasting. *Blasting* 4: 63–69.

Zhang, L.M. 1992b. *Models and applications of artificial neural network*. Shanghai: Fudan University Press, pp. 43–60.

Zhang, X.T. & Chen, S.H. 2002. Study on blast fragmentation for jointed and fractured rockmass considering collision. *Chinese Journal of Rock Mechanics and Engineering* 21(8): 1141–1146.

Zhou, C.B. 2003. Model of predicting the blasting fragmentation based on regression analysis. *Blasting* 20(4): 1–4.

Measurement and Analysis of Blast Fragmentation – Sanchidrián & Singh (Eds)
© 2013 Taylor & Francis Group, London, ISBN 978-0-415-62140-3

Assessing the effect of rock mass properties on rock fragmentation

A.K. Sirveiya
BSP, Steel Authority of India, Bhilai, India

N.R. Thote
National Institute of Technology, Nagpur, India

ABSTRACT: The strength of rock mass decreases with the increase in frequency of joints and the deformability of rocks depend on their orientation. Optimal blasting at all is considered to be based on the results of estimation of the rock mass properties and designation of the blasting method relating to them. The size distribution of a muck pile depends not only on the blasting standard but also on rock parameters like physico-mechanical properties, joints, rock mass characteristics etc. Information on the degree and size distribution of fragments within a blasted rock mass is essential for efficient rock loading and crushing operations. This study presents the result of a series of experiments carried out to evaluate the effects of rock mass properties on rock fragmentation. Ten experimental blasts were conducted at Iron ore mines and eight blasts were conducted at limestone mines exclusively to determine the effect of rock mass properties on rock fragmentation. WipJoint and WipFrag software used for joint and fragmentation analysis respectively.

1 INTRODUCTION

The simplest possible description of the rock mass for blasting purposes must say something about the intact rock strength and the structure of the rock mass. Blasting will loosen existing rock mass structure to liberate in-situ blocks as well as creating new fractures within the intact material. In the field of blasting technology the researchers are confronted with the problem of developing adequately accurate quantity; difficulties are to be the greatest part caused by the fact that the rock is neither homogeneous nor isotropic, the structural properties in the rock mass may, even when the rock type is the same, change from one site to another. A dominant influence on the results of blasting is exercised by the jointing system of the rock. Sources of geological data include exploration core logging and testing, surface and borehole geophysics, bench mapping and sampling, and equipment performance monitoring (Lizotte & Scoble 1994). Due to the heterogeneous nature of the rock mass, it is important to quantify the variability of these parameters.

Optical methods of analyzing fragmentation were first proposed by Carlsson & Nyberg (1983), and developed into a workable methodology by Maerz et al. (1987a, b). Since then, a proliferation of measurement systems have been described in the literature, including research tools (Nie & Rustan 1987, Paley et al. 1990, Doucet & Lizotte 1992, Stephansson et al. 1992, Montoro & Gonzalez 1993, Haverman & Vogt 1996, Girdner et al. 1996) and commercially available systems (Palangio et al. 1995, Maerz et al. 1996, Dahlhielm 1996, Schleifer & Tessier 1996, Downs & Kettunen 1996, Kleine & Cameron 1996, Chung & Noy 1996). Franklin et al. (1996) have given a brief history of the evolution of measuring systems. Whereas the standard of measurement is currently defined by sieving, optical methods are inevitably compared to sieving. The advantages of optical systems are numerous:

1. The measurements can be completely automated, eliminating the expense of a human operator, and the associated subjectivity.
2. Many more measurements can be made, consequently increasing statistical reliability by reducing sampling errors.
3. No interruption of production processes is required, and results are available in a very short time, allowing adjustments to production methods.
4. Screening is just too prohibitive in the case of large assemblages of rock or in the case of applications requiring very large blocks such as the evaluation of armorstone.

Recently, optical methods have come under criticism for proported lack of accuracy, inability to measure fines, and other various perceived deficiencies (Cunningham 1996). In part these criticisms are justified, under some conditions optical methods achieve very poor results. However there are many applications where results have justified the use of optical systems. Published applications using the WipFrag system include the following:

1. Selbay Mine, Joutel, Quebec, Canada, has optimized their blasting performance, monitoring energy consumption, loading rates, payloads of haulage trucks, secondary blasting costs, and maintenance costs as a function of fragment size (Palangio et al. 1995).
2. INCO's Coleman Mine, Sudbury, Ontario, Canada, was able to expand their blasting pattern by 40%, with cost saving of up to 80%, while actually improving the degree of fragmentation (Palangio et al. 1995).
3. Highland Valley Copper, Logan Lake, British Columbia, Canada, have been able to correlate their mill tonnage with the feed size, paving the way to greater production by optimizing feed size (Simkus and Dance 1998).
4. Bartley and Trousselle (1998) were able to show a direct relationship between accurate detonators and improved fragmentation.

2 FIELD INVESTIGATION

Six mines have been selected for this study. Total eighteen experimental blasts were conducted; out of these, ten blasts were conducted at four iron mines and eight blasts at two limestone mines. Rock samples were collected from each face to determine physico-mechanical properties of rock.

Experimental work comprises field study and laboratory work. Experimental blasts were conducted in iron ore and limestone mines situated in Jharkhand and Madhya Pradesh respectively. Laboratory work mainly comprises fragmentation analysis of each of the blast by using WipFrag system. The proposed methodology of evaluating the fragmentation efficiency is essentially based on the comparison of the block size distributions of the in-situ rock mass and corresponding muck pile.

The methodology is comprised of three stages.

1. Characterization of the rock mass, with the aim in view of characterizing the rock matrix properties and obtaining a quantitative estimate of the in-situ block size distribution.
2. Estimate of the muck pile fragmentation, likewise with the aim of obtaining a quantitative estimate of the block size distribution.
3. Evaluation of the results accomplished by the comparison of the block size distribution of the in-situ rock mass and those of the corresponding muck pile.

2.1 *Assessment of in-situ block size by using WipJoint system*

A better understanding of blasting requires new tools for describing the geological condition of the rock mass and the mechanical process that results in breakage. The in-situ block geometry and size distribution is calculated by WipJoint software to analyze the discontinuities. The blocks are transformed by the explosives energy into blasted rock fragments. Discontinuities such as bedding planes and joints slice the rock mass into blocks. Comminution theories can be applied to predict the size distributions of blast products. The volumetric joint count (J_v) is a simple measure of the degree of jointing or the in-situ block size of a rock mass. The number of joints intersecting one unit of rock mass is defined as the volumetric count and can be estimated from a joint count made within a window defined on the face.

Joint pattern analysis was performed with WipJoint software of the M/s WipWare Inc. The output of the analyses are in the form of number of exposed fragmented blocks, maximum, minimum and mean size of the fragmented blocks, sieve analysis as per the requirement i.e. at different percentile size viz. D_{10}, D_{25}, D_{50}, D_{75} and D_{90} (percentile sizes: for example D_{10} is the ten-percentile, the value for which 10% by weight of the sample is finer and 90% coarser; in terms of sieving, D_{10} is the size of sieve opening through which 10% by weight of the sample would pass), X_c is characteristic size, the intercept of the Rosin-Rammler straight line fitted to the WipFrag to data in log-log coordinates. This is equivalent to the $D_{63.2}$.

Tables 1 and 2 give the details of in-situ block size distribution of iron ore mine and limestone mine respectively.

2.2 *Assessment of fragmentation of production blasts by using WipFrag system*

After a blast it is essential to know its output especially in the form of fragmentation. Fragmentation can be evaluated by various methods such as analytical, observational and image analysis or photographic. Detailed analysis can be done by evaluating parameters like determination of exponent of uniformity, specific particle size etc. In this study fragmentation analyses were carried out with the help of WipFrag software. The output of the analyses are in the form of number of exposed fragmented blocks, maximum, minimum and mean size of the fragmented blocks, sieve analysis as per the requirement i.e. at different percentile size viz. D_{10}, D_{25}, D_{50}, D_{75} & D_{90}. Tables 3 and 4 give the details of blasted block size distribution of iron ore mine and limestone mine respectively.

Table 1. In situ size distribution of iron ore mine (m).

S. no.	Mine	D_{10}	D_{25}	D_{50}	D_{75}	D_{90}	X_{max}	X_c	n
1	A1	0.19	0.30	0.51	1.04	1.44	0.96	0.73	1.72
2	A2	0.33	0.47	0.70	1.10	1.81	1.20	0.88	1.52
3	A3	0.26	0.44	0.69	1.15	1.85	1.21	0.88	1.46
4	A4	0.11	0.24	0.39	0.59	0.87	0.65	0.46	1.35
5	B1	0.18	0.25	0.39	0.54	0.66	0.56	0.47	2.75
6	B2	0.13	0.24	0.47	0.63	0.83	0.61	0.68	2.14
7	C1	0.15	0.19	0.29	0.48	0.68	0.47	0.41	1.56
8	C2	0.18	0.27	0.36	0.52	0.73	0.53	0.47	1.65
9	D1	0.26	0.49	0.74	1.03	1.19	1.10	0.88	2.08
10	D2	0.23	0.36	0.62	0.95	1.23	0.98	0.84	1.85

Table 2. In situ block size distribution of limestone mine (m).

S. no.	Mine	D_{10}	D_{25}	D_{50}	D_{75}	D_{90}	X_{max}	X_c	n
1	A1	0.17	0.29	0.50	0.82	0.96	0.88	0.68	1.80
2	A2	0.97	0.14	0.20	0.32	0.45	0.32	0.25	2.01
3	A3	0.16	0.21	0.30	0.43	0.54	0.43	0.36	2.56
4	A4	0.10	0.15	0.33	0.54	1.03	0.69	0.43	0.97
5	B1	0.11	0.14	0.24	0.36	0.51	0.42	0.29	1.30
6	B2	0.10	0.21	0.39	0.54	0.74	0.52	0.59	1.62
7	C1	0.11	0.16	0.24	0.33	0.46	0.34	0.27	2.22
8	C2	0.16	0.24	0.33	0.46	0.66	0.47	0.41	1.95
9	D1	0.22	0.34	0.55	0.81	0.96	0.84	0.69	2.31
10	D2	0.21	0.33	0.52	0.78	1.12	0.84	0.71	2.13

Table 3. Blasted block size distribution of iron ore mine (m).

S. no.	Mine	D_{10}	D_{25}	D_{50}	D_{75}	D_{90}	X_{max}	X_c	n
1	E1	0.17	0.41	0.97	1.40	2.47	1.56	1.16	1.51
2	E2	0.24	0.42	0.93	2.21	2.76	1.20	1.46	1.47
3	E3	0.27	0.38	0.65	1.32	1.94	1.27	0.99	1.85
4	E4	0.39	0.70	1.37	2.42	3.13	2.46	1.87	1.67
5	F1	0.34	0.53	0.90	1.40	1.79	1.43	1.15	2.04
6	F2	0.28	0.46	0.94	2.22	2.92	2.01	1.46	1.42
7	F3	0.26	0.35	0.56	1.00	1.45	0.91	0.72	1.91
8	F4	0.24	0.31	0.47	0.95	1.23	0.85	0.74	1.84

Table 4. Blasted block size distribution of limestone mine (m).

S. no.	Mine	D_{10}	D_{25}	D_{50}	D_{75}	D_{90}	X_{max}	X_c	n
1	E1	0.20	0.29	0.49	0.93	1.36	0.77	0.60	1.83
2	E2	0.17	0.26	0.45	0.87	1.39	0.78	0.60	1.70
3	E3	0.11	0.19	0.42	0.78	1.18	0.84	0.59	1.26
4	E4	0.11	0.18	0.56	0.95	1.15	1.16	0.81	1.27
5	F1	0.10	0.16	0.27	0.46	0.79	0.50	0.39	1.73
6	F2	0.18	0.35	0.66	1.14	1.59	1.23	0.88	1.36
7	F3	0.12	0.18	0.28	0.42	0.60	0.42	0.34	2.27
8	F4	0.19	0.24	0.37	0.83	1.08	0.73	0.64	2.13

2.3 *Comparison of in-situ block size to blasted block size*

The prediction of the In-Situ Block Size Distribution (IBSD) was presented as a vital first step towards better prediction of the Blasted Block Size Distribution (BBSD), commonly termed the yield curve or fragmentation curve. The way in which in-situ bedding, jointing and other discontinuities slice up the natural rock mass into blocks of predefined shape distributions and size distributions prior to blasting is illustrated in virtually every exposure of rock. The quantification of the percentages of blocks bounded by joints or bedding planes within a rock mass was done. The concentrated release of energy from explosives detonated in confined blastholes, transforms the IBSD to a BBSD of finer material. Amongst all other physico-mechanical properties of rocks, compressive strength and tensile strengths are more important as per the rock fracture theories. The detailed investigations were done during experimentation and graphs were plotted against the median fragment size (D_{50}) of different blasts. The trend varies mines to mines but in general no significant trend has been established. This may be because of limitation on field studies or interferences of various variable parameters. The in-situ block size and blasted block size is calculated for each blast. Figures 1 to 7 illustrate

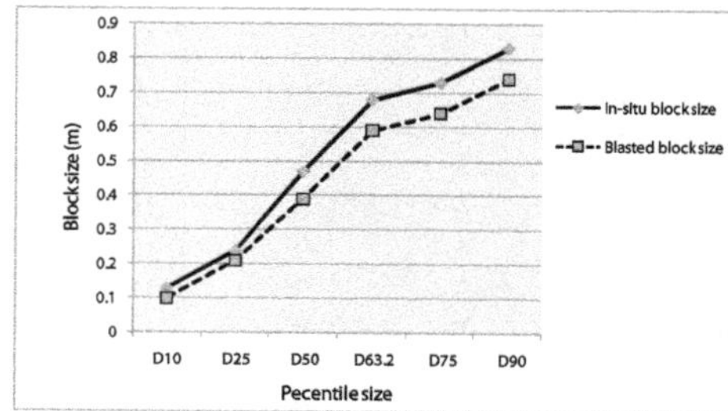

Figure 3. In-situ and blasted block size distribution of blast B1.

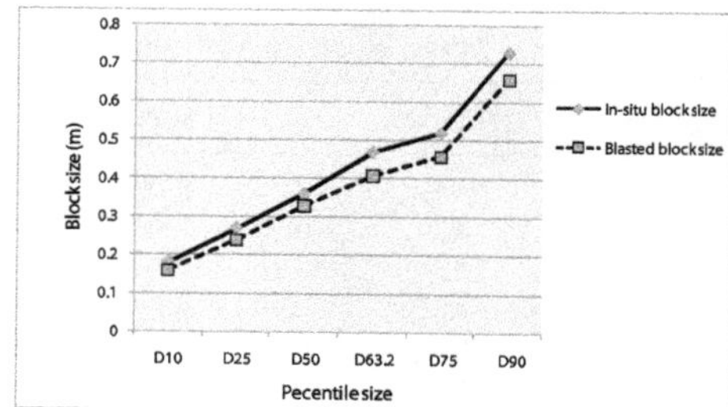

Figure 4. In-situ and blasted block size distribution of blast C1.

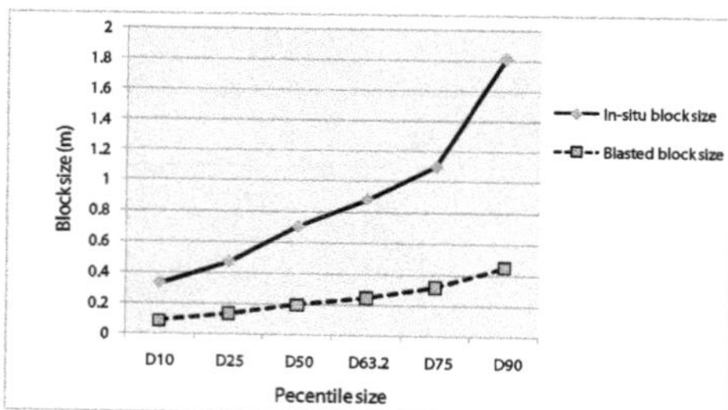

Figure 1. In-situ and blasted block size distribution of blast A1.

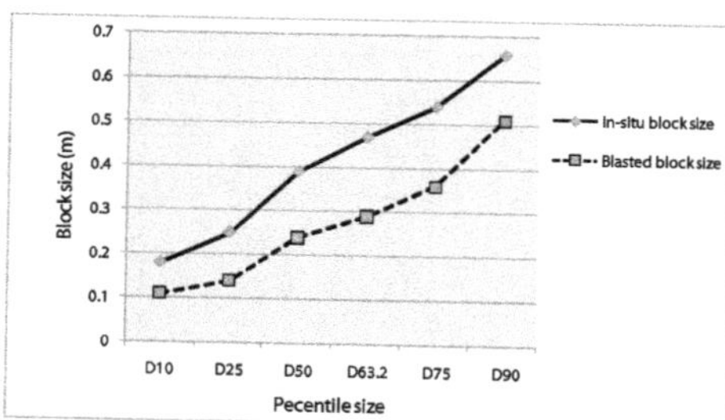

Figure 2. In-situ and blasted block size distribution of blast A4.

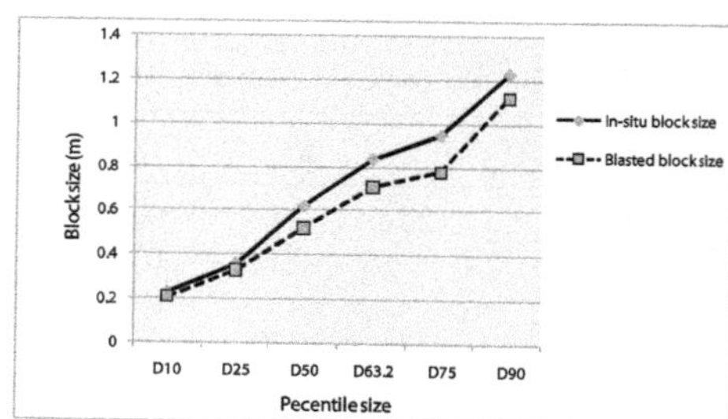

Figure 5. In-situ and blasted block size distribution of blast D1.

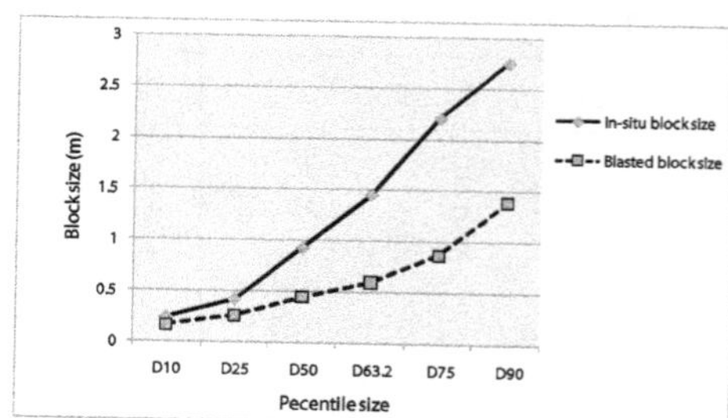

Figure 6. In-situ and blasted block size distribution of blast E1.

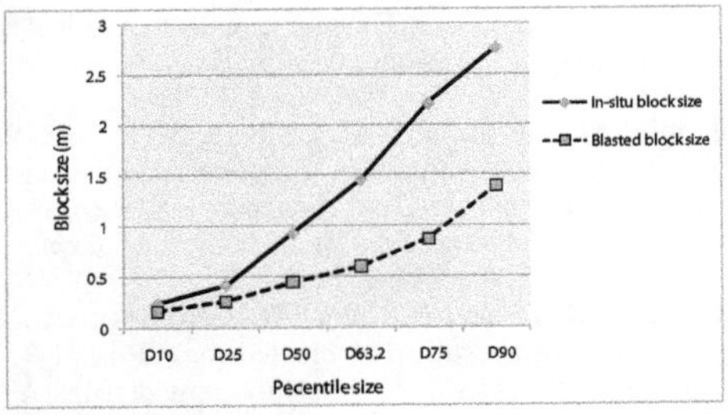

Figure 7. In-situ and blasted block size distribution of blast F1.

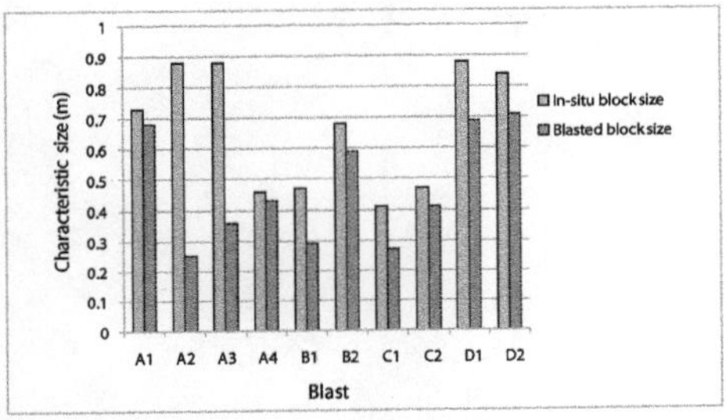

Figure 8. In-situ & blasted block characteristic size in iron ore mines.

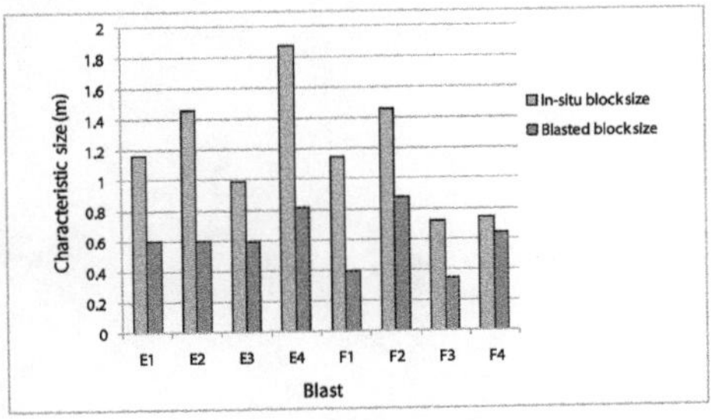

Figure 9. In-situ & blasted block characteristic size in limestone mines.

the comparison between in-situ block size distribution and blasted block size distribution for blasts A1 to F1. Figures 8 and 9 show the in-situ and blasted block characteristic size of iron ore mines and limestone mines respectively.

3 CONCLUSIONS

The following conclusions may be derived:

- Reduction in in-situ block size to blasted block size varies and depends on energy utilization during blasting.
- Fragmentation for the smaller blocks size is less than that of larger block size.
- There is more reduction in in-situ block size due to blasting at limestone mines than iron ore mines.

REFERENCES

Bartley, D.A. & Trousselle, R. 1998. Daveytronic, digital detonator testing in a vibration sensitive environment. *Proc. 24th Ann. Conf. on Explosives and Blasting Technique, New Orleans, Louisiana, USA*, pp. 247–261.

Carlsson, O. & Nyberg, L. 1983. A method for estimation of fragment size distribution with automatic image processing. *Proc. First Int. Symp. on Rock Fragmentation by Blasting, Luleå, Sweden*, pp. 333–345.

Chung, S.H. & Noy, M.J. 1996. Experience in fragmentation control. Proc. FRAGBLAST 5 Workshop on Measurement of Blast Fragmentation, Montreal, Quebec, Canada, 23–24 Aug. 1996, pp. 247–252.

Cunningham, C.V.B. 1996. Optical fragmentation assessment—A technical challenge. Proc. FRAGBLAST 5 Workshop on Measurement of Blast Fragmentation, Montreal, Quebec, Canada, 23–24 Aug. 1996, pp. 13–19.

Dahlhielm, S. 1996. Industrial applications of image analysis—The IPACS system. Proc. FRAGBLAST 5 Workshop on Measurement of Blast Fragmentation, Montreal, Quebec, Canada, 23–24 Aug. 1996, pp. 59–65.

Doucet, C. & Lizotte, Y. 1992. Rock fragmentation assessment by digital potography analysis. CANMET, MRL 92-116.

Downs, D.C. & Kettunen, B.E. 1996. On-line fragmentation measurment utilizing the CIAS system. *Proc. FRAGBLAST 5 Workshop on Measurement of Blast Fragmentation, Montreal, Quebec, Canada, 23–24 Aug. 1996*, pp. 79–82.

Franklin, J.A., Kemeny, J.M. & Girdner, K.K. 1996. Evolution of measurement systems: A review. *Proc. FRAGBLAST 5 Workshop on Measurement of Blast Fragmentation, Montreal, Quebec, Canada, 23–24 Aug. 1996*, pp. 47–52.

Girdner, K.K., Kemeny, J.M., Srikant, A. & McGill, R. 1996. The split system for analyzing the size distribution of fragmented rock. *Proc. FRAGBLAST 5 Workshop on Measurement of Blast Fragmentation, Montreal, Quebec, Canada, 23–24 Aug. 1996*, pp. 101–108.

Haverman, T. & Vogt, W. 1996. A system for the estimation of fragmentation after production blasts. *Proc. FRAGBLAST 5 Workshop on Measurement of Blast Fragmentation, Montreal, Quebec, Canada, 23–24 Aug. 1996*, pp. 67–71.

Kleine, T.H. & Cameron, A.R. 1996. Blast Fragmentation measurement using GoldSize. Proc. FRAGBLAST 5 Workshop on Measurement of Blast Fragmentation, Montreal, Quebec, Canada, 23–24 Aug. 1996, pp. 83–89.

Lizotte, Y.C., & Scoble, M.J. 1994. Geological control over blast fragmentation, *Can. Inst. Min. Met. Bulletin* 87: 57–71.

Maerz, N.H., Franklin, J.A., Rothenburg, L. & Coursen, D.L. 1987a. Measurement of rock fragmentation by digital photoanalysis. *ISRM. 6th Int. Cong. on Rock Mechanics, Montreal, Canada*, v1, pp. 687–692.

Maerz, N.H., Franklin, J.A. & Coursen, D.L. 1987b. Fragmentation measurement for experimental blasting in Virginia. *Proc. 3rd Mini-Symposium on Explosives and Blasting Research, Miami, Feb. 5–6*, pp. 56–70.

Maerz, N.H., Palangio, T.C. & Franklin, J.A. 1996. WipFrag image based granulometry system. Proc. FRAGBLAST 5 Workshop on Measurement of Blast Fragmentation, Montreal, Quebec, Canada, 23–24 Aug. 1996, pp. 91–99.

Montoro, J.J. & Gonzalez, E. 1993. New analytical techniques to evaluate fragmentation based on image analysis by computer methods. *Rock Fragmentation by Blasting*: 309–316. Rotterdam: Balkema.

Nie, S. & Rustan, A. 1987. Techniques and procedures in analysing fragmentation after blasting by photographic method. *Proc. 2nd Int. Symp. on Rock Fragmentation by Blasting, Keystone, USA*, pp. 102–113.

Paley, N., Lyman, G.L. & Kavetsky, A. 1990. Optical blast fragmentation assessment. *Fragblast '90, Brisbane, Australia, 26–31 Aug, 1990*, pp. 291–301.

Palangio, T.C., Franklin, J.A. & Maerz, N.H. 1995. WipFrag—A Breakthrough in Fragmentation Measurement. *Proc. 6th High-Tech Seminar on State of the Art Blasting Technology, Instrumentation, and Explosives, Boston, Mass.*, pp. 943–971.

Schleifer, J. & Tessier, B. 1996. FRAGSCAN: A tool to measure fragmentation of blasted rock. Proc. FRAGBLAST 5 Workshop on Measurement of Blast Fragmentation, Montreal, Quebec, Canada, 23–24 Aug. 1996, pp. 73–78.

Stephansson, O., Wang, W.X. & Dahlhielm, S. 1992. Automatic image processing of aggregates. *Proc. ISRM Symposium: Eurock '92, Chester, UK*, 14–17 September 1992, pp. 31–35.

Measurement and Analysis of Blast Fragmentation – Sanchidrián & Singh (Eds)
© 2013 Taylor & Francis Group, London, ISBN 978-0-415-62140-3

Author index

For Product Safety Concerns and Information please contact our EU
representative GPSR@taylorandfrancis.com
Taylor & Francis Verlag GmbH, Kaufingerstraße 24, 80331 München, Germany

www.ingramcontent.com/pod-product-compliance
Lightning Source LLC
Chambersburg PA
CBHW081210180926
51613CB00003B/198
* 9 7 8 0 4 1 5 6 2 1 4 0 3 *